Protecting Yourself and Your Family Against
Radiation Toxicity

Dr. Sundardas D. Annamalay

Published by Motivational Press, Inc.
7777 N Wickham Rd, # 12-247
Melbourne, FL 32940
www.MotivationalPress.com

Copyright 2014 © Sundardas D. Annamalay
All Rights Reserved

www.AgainstRadiation.com

No part of this book may be reproduced or transmitted in any form by any means: graphic, electronic, or mechanical, including photocopying, recording, taping or by any information storage or retrieval system without permission, in writing, from the authors, except for the inclusion of brief quotations in a review, article, book, or academic paper. The authors and publisher of this book and the associated materials have used their best efforts in preparing this material. The authors and publisher make no representations or warranties with respect to accuracy, applicability, fitness or completeness of the contents of this material. They disclaim any warranties expressed or implied, merchantability, or fitness for any particular purpose. The authors and publisher shall in no event be held liable for any loss or other damages, including but not limited to special, incidental, consequential, or other damages. If you have any questions or concerns, the advice of a competent professional should be sought.

Manufactured in the United States of America.

ISBN: 978-1-62865-087-7

CONTENTS

INTRODUCTION 7

1. Is There A Wireless Conspiracy? 9
2. Are We Microwaving Our Brains? 24
3. What Harmful Radiation Bombards Us Daily? 43
4. How Radiation Damages Our Cells 67
5. Calcium And Radiation Toxicity 82
6. What Secrets Does the US Army Know About Radiation? 102
7. How Do We Keep Ourselves And Our Families
   Safe From Radiation? 120

Appendix 1: Threshold Levels and Standards for
   Non-Ionising Radiation 138
Appendix 2: Basics of Radiation Theory 143
Appendix 3: Cellular Water and Radiation 155

TABLE OF ILLUSTRATIONS

FIGURES

Fig. 1: Life Blood Cells assessment and Electrosmog
Fig. 2: Power Absorbed by the Human Body
Fig. 3: The earth's natural magnetic field
Fig. 4: Exponential growth in background non-ionizing radiation
Fig. 5: Man-made versus natural EMFs.
Fig. 6: Once we have many cells communicating in a living tissue, they may act cooperatively to amplify a faint electromagnetic sign
Fig. 7: Relative strengths of the various biomagnetic fields measured in the spaces around the human body (Williamson & Kaufman, 1981).
Fig. 8a: A normal resting cell: a cell with a uniform distribution of charges surrounding the negatively (−) charged membrane. (left)
Fig. 8b: A cell influenced by EMFs: a cell both with negative charges in the membrane and positive charges concentrating in the direction of the exogenous field.(right)
Fig. 9: EMF induced biological effects
Fig. 10: Summary of scientific findings of ELF, RF/MW fields on cells
Fig. 11: Biochemical/biological/physiological events conditions and potential disease
Fig. 12: How the EMF technology blocks biological effects
Fig. 13: The time variance of a natural EMF frequency, amplitude and waveform varies at random (top). A "man-made" EMF, emitted from the internal circuitry of a digital mobile telephone. Frequency, amplitude and waveform are constant for a certain time period (bottom).
Fig. 14: Noise field superimposed on the constant ELF emitted from the circuitry of digital mobile phone
Fig. 15: EM noise inhibits the 60-Hz ELF-induced ODC activity in cells
Fig 16: Differences in Electromagnetic Field
Fig. 17: Impact of artificial radiation on cell wall
Fig 18: The Wave Rider

TABLES

Table 1: International Exposure Standards and Guidelines
Table 2: Health concerns regarding Safety Guidelines
Table 3: ELF safety standards.
Table 4: Typical ELF levels in mG (50/60 Hz) emitted by appliances and devices

INTRODUCTION

Your cellular phone sends out energy or radio waves that are very similar to the ones used by your microwave oven to reheat leftover pizza. Does that make you wonder what your cell phone might be doing to your brain?

Some internet sites claim that cell phones are so dangerous and powerful you can use their radiation to pop corn, boil an egg and even roast a chicken. That would be scary if it were true, but it's NOT true.

It's also not true that the electromagnetic fields created by cell phones are perfectly safe. That's what the phone manufacturers would like us to believe. They've commissioned studies that conveniently couldn't find any increased health risks at all from cell phone use.

The truth lies somewhere in the middle. Here's what you need to know about the risk of using your cell phone and a few simple ways you can protect yourself.

DO CELL PHONES REALLY COOK YOUR BRAIN?

In our modern world, we're bombarded by so many electromagnetic waves it's been called "electromagnetic pollution." It comes from appliances like hairdryers, refrigerators, computers, microwave ovens, air conditioners, compact fluorescent bulbs, vacuum cleaners and, of course, cell phones. Most researchers say these radio waves are weak and harmless to humans. Is this really true?

But cells phones present a special case. For one thing, we spend more time using them every day than other appliances and we use them right next to our brains. What's more, cell phone antennae from cell phone towers have stronger electromagnetic fields than other electronic gadgets.

From scientific studies, we definitely know that our brains can absorb radio waves from cell phones, and cell phone use can raise our brain's temperature by 7 degrees Fahrenheit. That's a lot!

A National Institutes for Health study also found that people who held a cell phone to their ear for more than 50 minutes had a 7% increase in their brain activity in the areas closest to the phone's antenna.

But does this increased heat and brain activity necessarily add up to brain damage? It has been proven, that over time these effects can lead to brain tumors.

Researchers studying large groups of people have found a clear association between cell phone use and an increase in brain tumors:

- In a 10-year study published in 2010 in the International Journal of Epidemiology, researchers reviewed the phone usage of over 22,000 people in 13 countries. They determined that cell phone users who spent 30 minutes or more per day on the phone had a 40% higher risk of developing a brain tumor on the side of the head where they usually held their phone.

- Just last year Italian researchers reviewed a large number of the studies on cell phones and found serious flaws in the research sponsored by cell phone companies. When they excluded phone industry research and relied only on independent third party studies, they concluded that cell phone use over a 10 year period doubles the risk of developing a brain tumor on the side of the head where the phone is habitually held.

In 2011, based on its own review of the studies, the World Health Organization (WHO) classified cell phone use as a possible cancer risk to humans. It's hard to ignore that kind of warning, so what can you do to protect yourself and your family? In Chapter 1 you will discover WHO recommendations to protect yourself against hand phone radiation. Read on to discover what you can do next.

1. IS THERE A WIRELESS CONSPIRACY?

THE UNITED STATES HAS 264 MILLION WIRELESS SUBSCRIBERS. THESE SUBSCRIBERS rely on two million cell towers and antennas to relay messages. Annually, subscribers relay 2.1 trillion phone minutes and 48 billion text messages. Cell phone towers are so ubiquitous they are now even a recognizable feature on the Himalayas.

Such a seismic global shift in technology occurring at such a quantum pace. Should we be concerned about health risks for children, our grandchildren and ourselves? Does the radio frequency exposure from thousands of minutes a month of cell phone calls pose an issue? Does the continuous exposure from millions of wireless networks, hot spots, cell towers, and antennas pose serious health problems?

> **What you need to know**
>
> The evidence that will unfold will reveal in no uncertain terms that we are bathed in a sea of toxic electromagnetic radiation (EMR). There has been a long and complicated chain of events leading to the proliferation of EMR devices. We have been cajoled, persuaded and seduced into believing that all these devices will make our life so much easier and are perfectly safe.
>
> As we delve deeper into the tangled roots of this global conspiracy, you will notice mention of the FDA cropping up. Whenever issues related to medical issues, health and money are involved, the FDA keeps popping up. As we will show later in the course of this book, by 1950s groups in quite a few countries as well as the US army were already concerned about the negative effects of EMR. The subsequent research did not allay these concerns satisfactorily. By the mid-1970s, ongoing research on the reproductive health of miniature swine, behavior of bees, bird development, growth of plants and the health of human beings was already demonstrating health challenges and safety issues connected to the use of EMR. We just never got to hear about them.

> The initial group of cell phone manufacturers pressured government regulatory agencies particularly the Food and Drug Administration (FDA) to allow cell phones to be sold without being tested for safety. As a result the FDA allowed cell phones to be sold without pre-market testing and without regulations.*
>
> In August 2007, an international group of scientists, researchers and public health policy professionals released a state-of-the-art report the "Bioinitiative Report 2007". This report was on the impact of exposure to harmful electromagnetic radiation (EMR) on our health.
>
> It documents scientific evidence showing the increased incidence of childhood leukemia, brain tumors, neurological impairment, leaking of blood brain barrier, learning disabilities and ADD. It also documents compromised immunity, hormone disruption. Alzheimer's disease, enzyme syntheses and genetic damage. It provides overwhelming evidence implicating electromagnetic radiation as a significant risk factor for both childhood and adult cancers. Information that we have not been made aware of.
>
> * "The Food and Drug Administration (FDA) says that available scientific evidence—... shows no increased health risk due to radiofrequency (RF) energy, a form of electromagnetic radiation that is emitted by cell phones" FDA website 2013

As we delve deeper into the tangled roots of this global conspiracy, you will notice the FDA keeps popping up. From the scientific validity of vaccination to the safety of radiation devices, the FDA has consistently supported monied interests with scant regards to the interests of the public. Let us look at how the foundations of this global conspiracy were laid in place historically.

ELECTROTHERAPY

Bioelectromagnetics as a field of study has it roots in the 18th and 19th centuries. This was when scientists in Europe and the United States first began to investigate the medical application of electromagnetism. Every American schoolchild can recite the story from 1752 when Benjamin Franklin flew a kite. The kite was flown with a key on the tail string during a thunderstorm. This was to show the equivalence of "electric fire" in lightning with "electric fire" created by rubbing specific objects together.

Electromagnetism, of course, can be traced to Thales and the ancient Greeks. However it was the work of 18th century European scientists and physicians that laid the groundwork for contemporary science.

Luigi Galvani (1737-1798), a Bolognese obstetrician and surgeon and Franklin's contemporary, used direct current electricity to treat tumors, aneurysms and fungal growths. The work of Faraday, Gilbert, Kelvin, Helmholtz, Maxwell and Hertz carried the study of the physical nature of electromagnetism through the 19th century. Perhaps the greatest contribution to the medical understanding of the effects of electromagnetism was made by Arséne d'Arsonval (1851-1940).

ELECTRICAL ACTIVITY MEASUREMENT SYSTEMS

The late 19th and the following 20th century brought rapid advances in the areas of biophysics and neurophysiology.

Early Researchers

- Galvani introduced the term "animal electricity",
- Alessandro Volta (1745 - 1827) invented the electric pile.
- The work of Faraday, Gilbert, Kelvin, Helmholtz, Maxwell and Hertz carried study of the physical nature ofelectromagnetism through the 19th century.
- Arséne d'Arsonval (1851-1940) invented diathermy and several measuring devices, including a thermocouple ammeter, and a moving-coil galvanometer, which became the basis for all pointer-type electric meters used subsequently.
- August Waller's 1887 discovery that the electrical activity of the human heart could be measured by the capillary electrometer.
- In 1906, Einthoven recorded the electricity from the heart using a sensitive galvanometer.
- In 1914 first electrocardiogram (EKG) machine made in the United States by Horatio Williams.

- German psychiatrist Hans Berger discovered the electroencephalogram (EEG) in 1929
- Duchene (1867) used fresh cadavers-places where nerves entered the muscles-electrical activity measured to see if muscle is properly enervated (electromyography).
- In 1942 Jasper discovered the electromyogram.

RADIO-FREQUENCY RESEARCH

Diathermy is the technique of using high-frequency energy to heat deep tissues in the human body. It was widely used by the medical community in Europe, North America and the Soviet Union during the first third of the 20th century. Research on RF energy absorption by Herman Schwan added understanding of the biophysical including many applications unrelated to diathermy. The basic research by Schwan and his colleagues produced many practical benefits. Commercial applications of microwaves burgeoned in the post-World War II era, particularly in the development of safety standards.

A parallel track of RF research in the years before World War II led to the development of radar in 1935. Radar was originally envisioned by its British inventor, Sir Robert Watson-Watt, as an aid to meteorological applications. Watson-Watt's vision was eventually fulfilled in the 1980s and 1990s. This was when the sophisticated new, NexRad Doppler weather radar was introduced. In the late 1930s, radar's application for aircraft detection was of paramount importance to Great Britain because of the rise of Nazi Germany.

Three streams of electromagnetic field research emerged in the 1940s and 1950s. First the military uses of microwaves, begun before World War II and was stimulated by the Cold War. Second there was a rapid expansion of electric power generation and transmission. Thirdly, there an increased interest in topics in zoology, physiology, biomedical investigation and applications.

The manner in which they would flow together as "bioelectromagnetics" might not have been evident at the time. However practical necessity and scientific curiosity drove researchers to ask related questions about the electrical and magnetic nature of living systems.

FORMATION OF THE BEMS

The eventual appearance of the Bioelectromagnetics Society (BEMS) had a strong driving force. There was the rapid increase in research on radio frequency technologies in the years after World War II. It was then scientists began questioning the safety levels of radiation exposure for humans. Especially so for those working with radar and radio-frequency devices used in medicine and industry.

Much of the technology was being adapted for military use. So the military services would originate many of the studies on the effects of exposure to EMR on military personnel.

During the Cold War period, the U.S. and U.S.S.R. scientists were often embarked on parallel research paths. It wasn't until the thaw in U.S.-Soviet relations during the 1970s that American scientists discovered the great volume of work done in the field by their Soviet counterparts.

In 1956, the U.S. Air Force and the Tri-Services ad hoc Committee became concerned about potential hazards of microwave radiation exposure to their personnel. Especially those working with the sophisticated new radar. The U.S. Air Force and the Tri-Services ad hoc Committee began a $13 million study to research biological effects at a wide range of frequencies and power levels. The research and the breakthrough results are examined in Chapter 6.

Civilian applications of new capabilities with radio frequency energy are exemplified by the microwave oven. This was first introduced by the Raytheon Corporation as the "Radarange" in 1947. It took 20 years before an affordably priced countertop model appeared as a consumer product with mass appeal. It would be another 20 years before we would discover the toxic effects of microwave cooking. By then the microwave oven had become an "essential" component in the modern kitchen and millions had been consuming microwaved Food.

Scientists, engineers and physicians working in the field soon made intriguing discoveries. Sol Michaelson, an early microwave researcher later active in BEMS, reported the breakdown of thermal regulation in a laboratory animal. This was brought about with a sufficiently high power level microwave radiation and long exposure time.

Some researchers reported that microwaves might interact with biological tissue without the production of heat. The stage was set for a decades-long argument between "thermalists" and "non-thermalists" over the existence of microwave-induced biological effects not caused by heating.

Programs run by Tom Rozzell and Elliot Postow provided much of the financial underpinning for the growing field of bioelectromagnetics in the 1970s. There was another major research program from the Air Force Office of Scientific Research (AFOSR) in Washington, D.C. This program was under Bill Berry, grant manager for the program. Other USA funding sources included NASA and the Bureau of Radiological Health (BRH).

Ross Adey at UCLA's Brain Research Institute received funding from AFOSR from the mid-1960s to the mid-1970s, as well as from NASA, ONR and BRH. In 1977, Eleanor Adair, then at Yale, applied to AFOSR for her first grant. She did basic research for them and her grant lasted for five years.

This work was conducted on squirrel monkeys. It measured how the monkeys regulate their body temperature both by behavior and through their autonomic physiological systems when they are irradiated with certain frequencies and intensities of RF. These early studies did not uncover any negative effects of RF.

STUDY ON BIOLOGICAL EFFECTS OF THE ELECTRIC UTILITY INDUSTRY

There was a rapid expansion in commercial nuclear power production in the late 1960s and 1970s. It occurred along with the corresponding increase in voltage of high voltage electric transmission lines. It led to establishing parallel fields of study in biological effects of transmission line fields at federal government agencies, quasi-independent research laboratories, and in the electric utility industry.

AEP, a midwestern power company with headquarters in the New York City of its birth, in the mid-1960s began funding studies in engineering, biophysics, human health, and biology. The purpose of these studies was to assess possible health effects on electric utility linemen.

These linemen were working on the new high capacity transmission lines that exposed them to strong electric and magnetic fields. In 1973, Chauncey Starr coordinated activities of the electric utility industry on a great many technical issues to create the Electric Power Research Institute (EPRI) in Palo Alto, California.

Among its first interests were the biological effects of electric power generation and transmission. By the mid- 1970s EPRI had a diverse program sponsoring studies such as those on the reproductive health of miniature swine, behavior of bees, bird development, and growth of plants.

Outside the United States, electric utilities sponsored much of the research on biological effects. In Canada, Ontario Hydro, Hydro-Québec and British Columbia Hydro were all active in funding studies. In Europe Électricité de France, the U.K.'s Central Electric Generating Board (CEGB), Japan's power industry, and several Swedish agencies were among the most active.

In Sweden, the power line issue started with the 1972 CIGRÉ meeting report on Russian switchyard workers. The debate that ensued became more energetic in light of plans for raising the voltage from 400 kV to 800kV on new power lines.

In the late 1970s, the Swedish Industrial Agency (Statens Industriverk), which later became the Board for Electrical Safety, in cooperation with the Swedish State Power Board and Sydkraft (Swedish power companies), financed a study. The study was on cows' fertility under power lines with first results appearing in a 1981 publication.

POSSIBLE TOXICITY EFFECTS OF ELECTRIC FIELDS AND AIR IONS

In the UK, the power-line issue was raised persistently in the mid 1970s on behalf of residents of the Dorset village of Fishpond Bottom. A 400-kV line passed low over the houses and sickness was apparently rife in that area. In 1978, a public inquiry was held into CEGB proposals to divert several spans of a 400-kV line closer to a school to make way for a new road. Concerns were expressed for the health effects of electric fields and air ions.

The inquiry spurred the CEGB's Central Electricity Research Laboratories (CERL) to initiate a programme of active research into possible effects of human exposure to strong power-frequency electric fields. At the time, research activity in Japan was limited to experiments on the effects of power line electric fields on plant growth.

These studies were conducted by scientists at the Central Research Institute of the Electric Power Industry (CRIEPI). Some experiments on the effects of electric fields on laboratory animals were conducted by an inter-institutional academic group in Sapporo. The Tokyo Electric Power (TEPCO) moved forward with plans for installation of a 1,100 kV line. This led to international reports about laboratory investigations and public concerns reaching Japan. As a result both government and industry increased research funding.

In addition, support from TEPCO enabled a wide ranging research program conducted by several universities and a toxicology institute. The program results were reported and summarized in a book published in both Japanese and English in 1999. At its beginning in 1980, bioelectromagnetics research in Spain was led by José M. Rodriguez-Delgado and Jocelyne Leal.

At that time, Delgado was the Director of Research of the world famous Hospital Ramóny Cajal in Madrid. He coordinated the work of a number of laboratories investigating mechanisms underlying therapeutic applications of electric and magnetic fields. Results of those early studies, which were supported by both public (Spanish NIH) and private (Electric Industry Consortium) funding. The results attracted international attention.

A study that reported developmental effects in embryonic chickens exposed to pulsed magnetic fields led to Project Henhouse. This was a productive international cooperative research project that emerged from discussions at the URSI Meeting held in 1984 in Florence, Italy.

Coordinated by the US Office of Naval Research and supported by US, Spanish, Swedish and Canadian public funds, Project Henhouse was an international attempt to reproduce the Delgado effects. Pulsed magnetic fields, however, continue to cause concern and controversy in the scientific community with respect to their potential detrimental effect on the development of chick embryos.

All the laboratories also examined the same endpoints. Exposed eggs were examined for fertility. Embryos were examined for abnormalities in development, and for growth. The "Henhouse Project" showed an overall increase in the proportion of abnormal embryos in the exposed group for all the laboratories taken together, although the exact proportion of abnormality differed from one laboratory to another.

By the late 1970s numerous research groups were studying dosimetry, biological effects and medical applications of EMF in France. The French Navy provided support for research on radiofrequency fields, mainly microwaves.

There is research that explains why the bees are particularly susceptible to microwave irradiation. The bees would be exposed to magnetic fields roughly 640 times more powerful than they normally encounter with the Earth's field. The consequences of this can be two-fold: i) the ferromagnetic compounds within their heads, thorax and abdomen can produce hysteresis loops affecting proprioception (spatial awareness); and ii) the very size of the bee's antennas, brain and body render it susceptible to resonance (unwanted vibrations). Put simply, I would argue that the bee is disorientated with a failing immune system and like AIDS in humans will become victim of any infection(s) or infestation(s) which came along.

BIOLOGICAL EFFECTS OF MILLIMETER WAVES

By the time of the 1980 meeting in Jouy-en-Josas, keen interest was generated by reports of biological effects of millimeter waves. Accompanying biophysical theories, and the potential for medical applications came from the Max-Planck-Institut für Festkörperforschung in Stuttgart, the Institute of Theoretical Physics at the University of Stuttgart, the Institute for Medical Engineering and Biophysics at the Forschungszentrum in Karlsruhe, the Gesellschaft für Strahlen- und Umweltforschung in Neuherberg, and the University of Tübingen.

The Federal Ministry for the Environment, Nature Conservation and Nuclear Safety (BMU) has supported research since 1990. The supported research was on the whole spectrum of non-ionizing radiation to address radiation protection for the public and also patients.

Since its founding in 1992, the Research Association for Radio Applications ("Forschungsgemeinschaft Funk", FGF) has supported research on high-frequency fields and communicated the results to the public.

In Italy, nationwide interest in the bioelectromagnetics issues was triggered in the mid 1970s by concerns about residential exposures to broadcasting emissions. A number of courses held in Erice, Sicily in the early 1980s gave a strong impulse to the development of studies in bioelectromagnetics.

In Italy, research funding came mostly from the Italian National Research Council, the Ministry of Health and the Ministry of Education, University, and Research, and other governmental Agencies. Canadian RF research of the late 1970s focused on microwave dosimetry and biological effects.

IS CELL PHONE RADIATION TOXIC?

George Carlo, PhD, JD, an epidemiologist and medical scientist who, from 1993 to 1999, headed the first telecommunications industry-backed studies into the dangers of cell phone use. That program remains the largest in the history of the issue. But he ran afoul of the very industry that hired him, when his work revealed preventable health hazards associated with cell phone use.

The cell phone industry is fully aware of the dangers. In fact, enough scientific evidence exists that some companies' service contracts prohibit suing the cell phone manufacturer or service provider, or joining a class action lawsuit. Still, the public is largely ignorant of the dangers. The media meanwhile, regularly trumpets new studies showing cell phones are completely safe to use. Yet, Dr. Carlo repeatedly pointed out, none of those studies could prove safety, regardless of who conducted them.

A WAKE-UP CALL

As I previously state, in August 2007, an international group of scientists, researchers and public health policy professionals released a state-of-the-art report "Bioinitiative Report 2007". This report was on the impact of electromagnetic fields (EMF) to our health It has since been updated to the "Bioinitiative Report 2012".

The BioInitiative 2012 Report has been prepared by 29 authors from ten countries, ten holding medical degrees (MDs), 21 PhDs, and three MSc, MA or MPHs. Among the authors are three former presidents of the Bioelectromagnetics Society, and five full members of BEMS.

It also documents compromised immunity, hormone disruption. Alzheimer's disease, enzyme syntheses and genetic damage. It provides overwhelming evidence implicating electromagnetic radiation as a significant risk factor for both childhood and adult cancers.

The research on the cell phones and brain cacer connection from Swedish brain tumor specialist Dr. Lennart Hardell, M.D., Ph.D. is pivotal in the debate about the safety of wireless radio frequency and microwave radiation.

"The evidence for risks from prolonged cell phone and cordless phone use is quite strong when you look at people who have used these devices for 10 years or longer, and when they are used mainly on one side of the head. Brain tumors normally take a long time to develop, on the order of 15 to 20 years. Use of a cell or cordless phone is linked to brain tumors and acoustic neuromas (tumor of the auditory nerve in the brain) and is showing up after only 10 years. A shorter time period than for most other known carcinogens," says Dr. Hardell.

More recently, in September 2008, Dr. Hardell told an international conference on radiation about mobile phone hazards. Those who started mobile phone use before the age of 20 had more than five-fold increase in glioma, a brain cancer. The extra risk to young people of contracting the disease from using the cordless phone was almost as great, at more than four times higher.

In March 2008, another powerful voice echoed this alarming trend; Dr. Vini Khurana, M.D., Ph.D., Associate Professor of Neurosurgery, Australian National University Medical School and Neurosurgeon at the Canberra Hospital, Australia. He released the findings of a 15-month "critical review" of the link between cell phones and malignant brain tumors and he too, concluded that using cell phones for more than 10 years could more than double the risk of brain cancer.

"It is anticipated that this danger has far broader public health ramifications than asbestos and smoking, and directly concerns all of us, particularly the younger generation, including very young children."

Most recently, Dr. Ronald B. Herberman, director of the University of Pittsburgh Cancer Institute issued an unprecedented warning about the possible health risks associated with cellular phone use, "Electromagnetic fields generated by cell phones should be considered a potential human health risk."

By1970's there were warnings from the animal studies that microwave radiation was not safe. Then the "Henhouse Project" showed an overall increase in the proportion of abnormal embryos due to microwave exposure. There was already considerable evidence that EMR from microwave and radiofrequency was not safe. Despite this, the cellular phone industry pressured government regulatory agencies—particularly the Food and Drug Administration (FDA)—to allow cell phones to be sold without pre-market testing. Some cell phone manufacturers also have special clauses that do not allow you to sue them.

Reynard's wife, Susan, died of a brain tumor, and he blamed cell phones for her death. So Florida businessman David Reynard filed a lawsuit against cell phone manufacturer NEC. The next day, telecommunications stocks took a big hit on Wall Street and the media had a field day. The industry trade association at the time, the Telecommunications Industry Association (TIA), went into crisis mode, claiming thousands of studies proved cell phones were safe and what Reynard and his attorney said was bunk.

TIA reassured the public that the government had approved cell phones, so that meant they were safe. The media demanded to see the studies, but, says Dr. Carlo, "The industry had lied. The only studies in existence then were on microwave ovens. At that time, 15 million people were using cell phones, a product that had never been tested for safety."

I trust I have laid bare the outlines of the conspiracy to inundate the world with wireless and toxic EMR devices, for the sake of profiteering. The public at large remains unaware of the conspiracy concocted by the cell phone industry. Meanwhile, concerned scientists are pointing out that EMR toxicity poses far greater health risks than even asbestos toxicity and smoking.

HOW TO PROTECT YOUR BRAIN FROM CELL PHONE RISKS

> The WHO's International Agency for Research on Cancer offers some practical tips to help reduce the risk of developing a brain tumor from your cell phone: (1997)
>
> - Try texting instead of calling when possible;
> - Use a hands-free device or the speaker mode to keep the phone's antenna away from your head;
> - Limit the number of calls you make from your cell phone and the length of time you talk;
> - Opt for a conventional land line when available; and
> - Avoid using your phone in areas with poor reception. If the signal is low, your phone has to put out more power to connect and that raises your risk.

REFERENCES

Abstract Book of the 17th Annual Meeting of BEMS, Boston, MA, June 18-22, 1995, The Bioelectromagnetics Society, Frederick, MD, 1995.

W. R. Adey: Frequency and Power Windowing in Tissue Interactions with Weak Electromagnetic Fields. Proceedings of the IEEE, 1980, vol. 63, no. 1, p. 119-125.

W. R. Adey: "Tissue interactions with nonionizing electromagnetic fields", Physiol. Rev., 1981, vol. 61, p: 435.

Hardell L, Carlberg M, Hansson Mild K. Pooled analysis of two case-control studies on the use of cellular and cordless telephones and the risk of benign brain tumours diagnosed during 1997-2003. Int J Oncol. 2006 Feb;28(2):509-18.

Hardell L, Mild KH, Carlberg M, Hallquist A. Cellular and cordless telephone use and the association with brain tumors in different age groups. Arch Environ Health. 2004 Mar;59(3):132-7.

Repacholi M and Greenebaum B (1998). Interaction of static and extremely low frequency electric and magnetic fields with living systems: health effects and research needs. Bioelectromagnetics (In press). (Summary report of WHO scientific review meeting on static and ELF held in Bologna, 1997

Hecht, K et al. 2007.Overloading of Towns and Cities with Radio Transmitters (Cellular Transmitter): a hazard for the human health and a disturbance of eco-ethics. IRCHET. International Research Centre of Healthy Ecological Technology.Berlin-Germany P.1, para. 3

NMRI. 1971.Biography of Reported Biological Phenonomena ('Effect') and Clinical Manifestations Attributed to Microwave and Radio-Frequency Radiation. Research Report. MF12.524.015-0004B Report No. 2. NMRI. National Naval Medical Centre. 4 Oct 1971

US Defence Intelligence Agency Documents.

DST-18105-076-76

DST-18105-074-76

ST-CS-01-169 -72

Thorsten Ritz et al.Resonance Effects Indicate a Radical-pair mechanism for Avian Magnetic Compass. Nature. Vol. 429. 13 May 2004. P. 177

Prof K Richter et al.Kompetenzinitiative. For the Protection of Man, Environment and Democracy. 16 March 2008 P. 3

Diagnose-Funk. The Big Bee Death, 4 April 2007 P.4

Guy Cramer. To Bee or Not To Bee, If That's the Question, What is the Answer? Colony Collapse Disorder linked to HAARP, 2 June 2007

Colin Buchanan. The Disappearing Bees: CCD and Electromagnetic Radiation, 22 February 2008

Barrie Trower. Presentation to the Beekeepers' Association Glastonbury 9 August 2008 'Is The Colony Collapse the price of e.m.f. progress?'

Joris Everaert et al.A Possible Effect of Electromagnetic Radiation from Mobile Phones Base Stations on the Number of Breeding House Sparrows (Passer Domesticus), Electromagnetic Biology and Medicine No. 26 pp. 63-72, 2007

Animal study - EMF Radiation.http://members.aol.com/gotemf/emf/animals.htm

Prof D Henshaw. So Much Research, Yet so Little Notice Taken Physics Department, Bristol University, UK

Microwave News.GSM Modulation is Key to Non Thermal, Neurological Effects. Vol. XXII No. 6 Nov / Dec. 2002 P.8

EMR and Plants: published papers in peer-reviewed scientific journals that show (possible) EMR effect. http://omega.twoday.net/stories/4601917

Robert Costanza et al. The Value of the World's Ecosystem Services and Natural Capital Nature. 387 pp 253-260 15 May 1997

G. Bennet.Powering Down the Networks 12 October 2009 http:totaltele.com/view.aspx?ID=449730

Bioinitiative. www.bioinitiative.org/report

2. ARE WE MICROWAVING OUR BRAINS?

IN THE PREVIOUS CHAPTER, I WROTE THAT DR. HARDELL HAD MENTIONED THAT THOSE who started mobile phone use before the age of 20 had more than five-fold increase in brain cancer Do you wonder if this was related to radiation hazards?

> **What you should know**
>
> The cellular phone industry was born in the early 1980s. This was when communications technology that had been developed for the Department of Defense was put into commerce by companies focusing on profits. This group, with big ideas but limited resources, pressured government regulatory agencies—particularly the Food and Drug Administration (FDA)—to allow cell phones to be sold without pre-market testing.
>
> The pressure worked, and cell phones were exempted from any type of regulatory oversight, an exemption that continues today. An eager public grabbed up the cell phones. without realizing the health consequences.
>
> The first standard adopted was derived largely from the work of the engineer Herman Schwan who assumed that heating was the only effect EMFs would have on living tissue. This led to totally inadequate safety levels for cell phone radiation.
>
> Scientists found that cell phone radiation caused DNA damage, impaired DNA repair, and interfered with cardiac pacemakers. European research confirmed Dr. Carlo's findings. Studies suggest that cell phone radiation contributes to brain dysfunction, tumors, and potentially to conditions such as autism, attention deficit disorder, neurodegenerative disease, and behavioral and psychological problems.
>
> In USA, SAR limit for cell phones is **1.6W/Kg** which is actually for **6 minutes.** Meanwhile Phillips et al. (1998) observed DNA damage at 0.024-0.0024 W/Kg. **Cellular damage was observed at levels one hundred to one thousand times below the USA safety level..** The safety levels for microwave radiation in New South Wales, Australia is 0.00001 W/m^2. Above these values symptoms occur.
>
> In India, ICNIRP recommends levels of 9.2 W/m^2, which is 100, 000 times more then

> the New South Wales (0.00001 W/m^2) levels. If you accept the **ICNIRP guidelines in India, you would be led to believe that you can keep your body in a microwave for 19 minutes**. This is obviously not true. The pulsed microwaves at low power was killing rats in a few seconds, evidently by inadvertent interference pattern "hot spots" of focused power burning holes in cell walls.
>
> **In other words non-thermal bioeffects and cellular damage occurs at almost one thousand to one hundred thousandth of the officially accepted safety levels.**
>
> I strongly believe that the safety levels recommended in New South Wales Australia are the safest for us in the long term. This in practice would make the use of hand phones unfeasible.

LAWSUIT PROMPTS SAFETY STUDIES

In 1993, the cell phone industry was pressured by Congress to invest $28 million into studying cell phone safety. The cause of this sudden concern was massive publicity about a lawsuit filed by Florida businessman David Reynard against cell phone manufacturer NEC.

Reynard's wife, Susan, died of a brain tumor, and he blamed cell phones for her death. Reynard revealed the suit to the public on the Larry King Live show, complete with dramatic x-rays showing the tumor close to where Susan held her cell phone to her head for hours each day.

CELL PHONES REACH THE MARKET WITHOUT SAFETY TESTING

The next day, telecommunications stocks took a big hit on Wall Street and the media had a field day. The industry trade association at the time, the Telecommunications Industry Association (TIA), went into crisis mode, claiming thousands of studies proved cell phones were safe and what Reynard and his attorney said was bunk.

The cellular phone industry was born in the early 1980s, when communications technology that had been developed for the Department of Defense was put into commerce by companies focusing on profits. This group pressured government regulatory agencies—particularly the Food and Drug Administration (FDA)—to allow cell phones to be sold without pre-market testing.

TIA reassured the public that the government had approved cell phones, so that meant they were safe. The media demanded to see the studies, but, says Dr. Carlo, "The industry had lied. The only studies in existence then were on microwave ovens. At that time, 15 million people were using cell phones, a product that had never been tested for safety."

Forced to take action, the cell phone industry set up a non-profit organization, Wireless Technology Research (WTR), to perform the study. Dr. Carlo developed the program outline and was asked to head the research.

Oversight of the issue was charged to the FDA, though it could have and probably should have gone to the Environmental Protection Agency (EPA), which fought hard for jurisdiction. But the industry had enough influence in Washington to get whatever overseer it wanted. It simply didn't want to tangle with EPA because, says Dr. Carlo, "… the EPA is tough."

"Anything that's ever made a difference in terms of public health has come from the EPA," he says. "But safety issues that are covered in corruption and questions seem to always have a connection to the FDA, which has been manipulated by pharmaceutical companies since it was born."

When called to help with the cell phone issue, Dr. Carlo was working with the FDA on silicone breast implant research. The choice of Dr. Carlo to head WTR seemed unusual to industry observers.

An epidemiologist whose expertise was in public health and how epidemic diseases affect the population, he appeared to lack any experience in researching the effects of EMR on human biology. Based on this, a premature conclusion was drawn by many: Dr. Carlo was an "expert" handpicked by the cell phone industry, and therefore his conclusions would only back up the industry's claim that cell phones are safe.

Dr. Carlo, however, refused to be an easy target. He quickly recruited a group of prominent scientists to work with him, bulletproof experts owning long lists of credentials and reputations that would negate any perception that the research was predestined to be a sham.

He also created a Peer Review Board chaired by Harvard University School of Public Health's Dr. John Graham, something that made FDA officials more comfortable since, at the time, the agency was making negative headlines due to the breast implant controversy. In total, more than 200 doctors and scientists were involved in the project.

STRICT STUDY GUIDELINES FOR STUDY DESIGNED BY DR CARLO

Once all involved agreed on what was to be done, Dr. Carlo presented the study's stakeholders in the industry, the government, and the public with a strict list of criteria for moving forward.

"The money had to be independent of the industry—they had to put the money in trust and couldn't control who got the funds," he says. "Second, everything had to be peer reviewed before it went public, so if we did find problems after peer review, we could use that information publicly to recommend interventions."

A third requirement was for the FDA to create a formal interagency working group to oversee the work and provide input. The purpose of this was to alleviate any perception that the industry was paying for a result, not for the research itself. But the fourth and last requirement was considered by Dr. Carlo to be highly critical: "Everything needed to be done in sunlight. The media had to have access to everything we did."

Dr. Carlo brought safety information about cell phones to the public through his book," Cell Phones: Invisible Hazards in the Wireless Age". The best protection against cell phone radiation is keeping a safe distance. Also to always use a headset to minimize exposure to harmful cell phone radiation.

New evidence was growing fast about health risks from mobile phones electromagnetic radiation. Dr Kjell Hansson Mild in Sweden studied radiation risk in 11,000 mobile telephone users. Symptoms such as fatigue, headaches, burning sensations on the skin were more common among those who made longer mobile phone calls.

At the same time there are a growing number of unconfirmed reports of individuals whose health has been affected after chronic, frequent use of mobile phones, presumably from radiation effects on cells. See Appendix 1 for SAR data on mobile phone radiation levels.

> In all, they conducted more than fifty studies that were peer-reviewed and published in a number of medical and scientific journals. Under Dr. Carlo's direction, scientists found that cell phone radiation caused DNA damage, impaired DNA repair, and interfered with cardiac pacemakers.
>
> European research confirmed Dr. Carlo's findings. These studies suggested that cell phone radiation contributes to brain dysfunction, tumors, and potentially to conditions such as autism, attention deficit disorder, neurodegenerative disease, and behavioral and psychological problems.
>
> Australian study: Mice exposed to pulsed digital mobile phone radiation over 18 months had twice the risk of developing cancers.
>
> American study: Learning and short term memory were impaired after 45 minutes exposure to electromagnetic radiation from mobile phones in rats. And other studies of electromagnetic radiation on pregnant mice suggest that high exposure to mobile phones can affect intra-uterine development, confirmed recently in chicks (double birth defects, Chapter 3 P.50). The effects of mobile phone radiation in human embryo development are unknown.
>
> UK government report: Clear that mobile phone radiation can affect brain function. Now that 20,000 radio masts in the UK are active it means that everyone is being subjected to constant low level electromagnetic radiation.

RELIABLE CELL PHONE RADIATION STUDIES

But what about longer term radiation effects of using mobile phones? Could mobile phone exposure trigger cancer? Birth defects? What about the health risks not just from mobile phones but the transmitter masts?

This work on human subjects follows other mobile phone studies in animals suggesting that electromagnetic radiation from mobiles may cause brain tumours, cancer, anxiety, memory loss and serious birth defects.

In Britain a 27 year old woman with a brain tumour is taking a mobile phone manufacturer to court who she blames for her tumour. A biologist, Roger Coghill has also been given permission to bring a case against a provider of mobile phone equipment for failing to warn people of radiation hazards.

A wide variety of electrical devices contribute to electrosmog, ranging from computers, to phones, TV sets, radar transmitter and transformers. However mobile phone radiation is certainly intense, as evidenced by the effects on aircraft navigation systems, or more obviously on a nearby conventional telephone or a music system.

New research suggests living downwind from an electricity pylon can increase the risk of lung cancer significantly. Ionisation of the air causes microscopic pollution particles to become charged so they stick to the lining of the lung.

FLAWED DANISH STUDY REPORTS CELL PHONES ARE SAFE

In December, 2006, an epidemiological study on cell phone dangers published in the Journal of the National Cancer Institute sent the media into a frenzy. Newspaper headlines blared: "Danish Study Shows Cell Phone Use is Safe," while TV newscasters proclaimed, "Go ahead and talk all you want—it's safe!"

The news seemed to be a holiday gift for cell phone users. But unfortunately, it's a flawed study, funded by the cell phone industry and designed to bring a positive result. The industry's public relations machine was working in overdrive to assure that the study get top-billing in the media worldwide.

According to Dr. George Carlo, the study, by its design, could not identify even a very large risk. Therefore, any claim that it proves there's no risk from cell phones is a blatant misrepresentation of the data that will give consumers a very dangerous false sense of security.

"Epidemiological studies are targets for fixing the outcome because they're observational in nature instead of experimental," Dr. Carlo explains. "It's possible to design studies with pre-determined outcomes that still fall within the range of acceptable science. Thus, even highly flawed epidemiological studies can be published in peer-reviewed journals because they're judged against a pragmatic set of standards that assume the highest integrity among the investigators."

The Danish Study was a flawed study funded by the cell phone industry.

WHY CELL PHONES ARE DANGEROUS

A cellular phone is basically a radio that sends signals on waves to a base station. The carrier signal generates two types of radiation fields: a near-field plume and a far-field plume. Living organisms, too, generate electromagnetic fields at the cellular, tissue, organ, and organism level; this is called the biofield.

Both the near-field and far-field plumes from cell phones and in the environment can wreak havoc with the human biofield, and when the biofield is compromised in any way, says Dr. Carlo, so is metabolism and physiology.

"The near field plume is the one we're most concerned with. This plume that's generated within five or six inches of the center of a cell phone's antenna is determined by the amount of power necessary to carry the signal to the base station," he explains. "The more power there is, the farther the plume radiates the dangerous information-carrying radio waves."

With the background levels of information-carrying radio waves dramatically increasing because of the widespread use of cell phones, Wi-Fi, and other wireless communication, the effects from the near and far-fields are very similar. Overall almost all of the acute and chronic symptoms seen in electrosensitive patients can be explained in some part by disrupted intercellular communication.

These symptoms of electrosensitivity include inability to sleep, general malaise, and headaches. Could this explain the increase in recent years of conditions such as attention-deficit hyperactivity disorder (ADHD), autism, and anxiety disorder?

Epidemic curve projections, however, indicate that in 2006, we can expect to see 40,000 to 50,000 cases of brain and eye cancer. This is based on published peer-reviewed studies that allow calculation of risk and construction of epidemic curves. By 2010, according to Dr. Carlo, we can expect that number to be between 400,000 and 500,000 new cases worldwide.

> **Here's how cell phone radiation causes damage:**
>
> A carrier wave oscillates at 1900 megahertz (MHz) in most phones, which is mostly invisible to our biological tissue and doesn't do damage. The information-carrying secondary wave necessary to interpret voice or data is the problem. That wave cycles in a hertz (Hz) range familiar to the body. Your heart, for example, beats at two cycles per second, or two Hz.
>
> Our bodies recognize the information-carrying wave as an "invader," setting in place protective biochemical reactions that alter physiology and cause biological problems that include intracellular free-radical buildup, leakage in the blood-brain barrier, genetic damage, disruption of intercellular communication, and an increase in the risk of tumors. The health dangers of recognizing the signal, therefore, aren't from direct damage, but rather are due to the biochemical responses in the cell.
>
> Cellular energy is now used for protection rather than metabolism. Cell membranes harden, keeping nutrients out and waste products in. Waste accumulating inside the cells creates a higher concentration of free radicals, leading to both disruption of DNA repair (micronuclei) and cellular dysfunction.
>
> Unwanted cell death occurs, releasing the micronuclei from the disrupted DNA repair into the fluid between cells (interstitial fluid), where they are free to replicate and proliferate. This is the most likely mechanism that contributes to cancer.
>
> Damage occurs to proteins on the cell membrane, resulting in disruption of intercellular communication. When cells can't communicate with each other, the result is impaired tissue, organ, and organism function. In the blood-brain barrier, for example, cells can't keep dangerous chemicals from reaching the brain tissue, which results in damage.

RADIO FREQUENCY OR MICROWAVE RADIATION

Radiation emitted from Cell Phones, Cell phone towers, Wi-Fi, TV and FM towers,

microwave ovens, etc. are sources of RF/MW. This type of EMR causes significant health hazards (biological effects) on human, animals, birds, plants and environment

The primary sources of RF/MW exposures are mobile phones, personal communication systems (PCS), mobile phone and PCS antenna towers, TV and radio broadcasting antennas, radar equipment, and two-way radios.

FIELD STRENGTH, SAFETY STANDARDS AND SOURCES OF MAN MADE RF FIELDS

Radio frequency (RF) and microwave (MW) field intensities are usually measured in milliwatts per square centimeter (mW/cm2). However, the intensity provides little information on the biological consequence unless the amount of energy absorbed by the irradiated object is known.

This is generally given as the specific absorption rate (SAR), which is the rate of energy absorbed by a unit mass (e.g., one kg or one g of tissue) of the object. The unit of measurement for the SAR is watts per kg (W/kg).

The rate of absorption and distribution of RF/MW energies depend on many factors like type and shape of tissue, orientation relative to the radiation, type and parameters of the radiation, etc. The distribution of absorbed energy in an irradiated organism is extremely complex and non-uniform, and may lead to the formation of so called "hot spots" of concentrated energy in the tissue.

Present US safety standards related to RF/MW EMF are based on thermal (heating) effects. The first standard adopted was derived largely from the work of the engineer Herman Schwan.

Safety Levels for microwave radiation based on thermal effects:	
1966 American National Standards Institute (ANSI) guidelines	10mW/cm^2
1980s Standards in the United Kingdom for short term exposure	25mW/cm^2
US Power density standards 800- 900 MHz cell phones	0.579mW/cm^2
US Power density standards PCS(public exposure)	1mW/cm^2
American Power density standards PCS(occupational exposure)	5mW/cm^2
US ANSI/IEEE standard for "uncontrolled" environments (in SAR)	1.6W/Kg
US ANSI/IEEE standard for for "controlled" environments(in SAR)	8W/Kg
European standard for "uncontrolled" environments (in SAR)	2W/kg
Japanese standard for "uncontrolled" environments (in SAR)	8W/kg

Although Schwan not a biologist, he assumed that heating was the only effect EMFs would have on living tissue. In the 1950s Schwan worked for the US Department of Defense, estimating danger levels based on how much radar MW energy was needed to measurably heat metal balls and beakers of salt water, which he used to represent the size and presumed electrical characteristics of various animals.

Appreciable heating occurred in these models only at levels of 100 mW/cm^2 or above, so, incorporating a safety factor of ten, Schwan in 1953 proposed an exposure limit of 10 mW/cm^2 for humans. No one ever tested for subtler effects, and the 10 mW/cm^2 level was uncritically accepted on an informal basis by industry and the military. Current research is showing that these levels are totally inadequate. There are non-thermal bioeffects and cellular damage occurring at almost one thousand to one hundred thousandth of the officially accepted safety levels.

Although it may be more relevant to use SAR, it is much more difficult to measure and control, requiring the use of "phantom-heads", models of human heads. Of course, these can never be identical to "real" human heads. Also for RF/MW exposures, research has demonstrated significant RF/MW induced effects on cells and animals at exposure intensities thousands of times below the various standard safety limits.

Non-thermal effects of cell phone radiation:
- The Danish scientists (1997), Kwee and Raskmark reported changes in human cell proliferation rates at SARs of 0.000021-0.0021 W/kg;
- Magras and Xenos (1997) reported a decrease in reproductive functions in mice exposed to RF/MW of intensity at 0.00016-0.001053 mW/cm^2;
- Ray and Behari (1990) reported decrease in eating and drinking behavior in rats exposed to 0.0317 W/kg;
- Dutta et al. (1989) reported changes in calcium metabolism in cells exposed to 0.05-0.005 W/kg;
- Phillips et al. (1998) observed DNA damage at 0.024-0.0024 W/Kg.

This means that the 1966 American National Standards Institute (ANSI) guidelines of 10mW/cm^2 totally inadequate. **Cellular damage occurs at one thousand to one hundred thousandth of the officially accepted safety levels.**

In December 1998, Dr. Hyland of the University of Warwick, UK, at the International Institute of Biophysics, Neuss-Holzheim, Germany, presented an attempt to draw attention to "a multitude of frequency-specific, non-thermal bioeffects – induced in living systems by ultra-low-intensity microwave radiation – the existence of which is not currently taken into account in the formulation of the safety limits to which microwave devices must conform".

Other scientists representing decades of practical experience within the field of RF/MW induced biological effects, Dr. Henry Lai of University of Washington, Seattle and Dr. Neil Cherry of Lincoln University, New Zealand, have found low-level effects and have reviewed the scientific literature, addressing a multitude of scientific evidence of low-level RF/MW induced bioeffects in cells and animals, neurological effects in humans and elevated risks of cancer in humans. The scientists setting the standard safety limits have chosen to disregard such evidence, showing that EMF induced effects may be cumulative, (DNA damage).

In other words non-thermal bioeffects and cellular damage occurs at almost one thousandth to then thousandth (0.00016-0.001053 mW/cm^2) of the official accepted safety levels (10mW/cm^2).

MICROWAVE HEATING CONCEPT

4.2 KW (4200 W) of microwave power raises temperature of 1 Litre of water by 10^{0}C in 1 second. In energy absorption term, 4.2 KW-sec microwave energy will increase the temperature of 1 Litre by 10^{0}C.

For example, in a microwave oven, temperature of one cup of water increases from 30^{0}C to 100^{0}C in approximately 70 seconds with 500W of microwave power. With 1W power (same as output power of cell phones), temperature will increase by 1^{0}C in 500 seconds.

Temperature of ear lobes increases by approximately 1^{0}C when cell phone is used for approximately 20 minutes. A cell phone transmits 1 to 2 Watts of power.

LIVE BLOOD CELLS & ELECTROSMOG

LIVE BLOOD: August 20 2009

SLIDE 1
Low EMF/EMR exposure [12:30 pm]

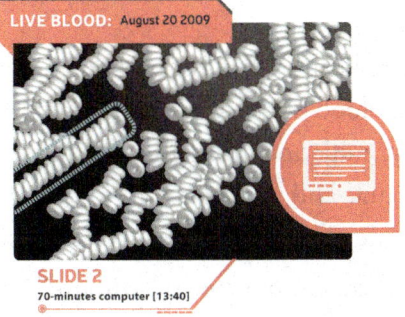

SLIDE 2
70-minutes computer [13:40]

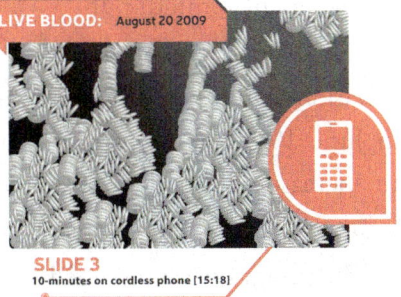

SLIDE 3
10-minutes on cordless phone [15:18]

Fig. 1: Life Blood Cells assessment and Electrosmog

Slide 1: Shows normal blood cell formation when exposed to low EMR. The blood cells are relatively not clumped together

Slide 2: Shows abnormal blood cell formation when exposed to higher levels of EMR. The blood cells are relatively clumped together. This is indicative of higher levels of free radical formation

Slide 3: Shows even higher levels of abnormal blood cell formation when exposed to even higher levels of EMR. The blood cells are even more clumped together. There is even more free radical formation and cellular damage.

SAR (Specific absorption rate) - Rate at which radiation is absorbed by human body, measured in units of watts per kg (W/kg) of tissue.

In USA, SAR limit for cell phones is 1.6W/Kg which is actually for 6 minutes. It has a safety margin of 3 to 4, so a person should not use cell phone for more than 18 to 24 minutes per day. This information is not commonly known to people. As a result, millions of Americans use the cell phone for hours a day, bombarding their brains and bodies with radiation that exceeds safety standards from a hundred thousand to a million times more.

ICNIRP has considered only thermal effects of radiation and has given following disclosure:

ICNIRP is only intended to protect the public against short term gross heating effects and NOT against 'biological' effects such as cancer and genetic damage from long term low level microwave exposure from mobile phones, masts and many other wireless devices.- http://ww.icnirp.de/documents/emfgdl.pdf

Now there is more evidence of definite health effects of mobile phone radiation on human health ranging from blood pressure to brain tumours. In June 1998 the Lancet reported that radiation from mobiles causes an increase of blood pressure. Dr Braune and colleagues in Freiburg, Germany, attached mobiles to the right side of the heads of ten volunteers.

The phones were switched on and off by remote control without the volunteers knowing - so that any radiation effect could be separated from the psychological effect of holding a mobile phone. Their blood pressure rose each time by between 5-10mm Hg, probably from an electromagnetic radiation induced constrictive effect on blood vessels from the mobile phones.

This level of increase would be more than enough to trigger a stroke or heart attack in someone at severe risk. This was the first firm evidence that mobile phone radiation could directly alter cell function in the human body.

2. ARE WE MICROWAVING OUR BRAINS?

International Exposure limits for RF fields (1800MHZ)	
9.2 W/m²	ICNIRP and EU recommendation 1998 - Adopted in India
3 W/m²	Exposure limit in Canada (Safety Code 6, 1997)
2 W/m²	Exposure limit in Australia
1.2 W/m²	Belgium (ex Wallonia)
0.5 W/m²	Exposure Limit in Auckland, New Zealand
.24 W/m²	Exposure limit in CSSR, Belgium Luxembourg
0.1 W/m²	Exposure limit in Poland, China, Italy, Paris,
0.095 W/m²	Exposure limit in Switzerl, Italy in areas with duration > 4hours
0.09 W/m²	ECOLOG 1998 (Germany) *Precaution recommendation only*
0.025 W/m²	Exposure limit in Italy in sensitive areas
0.02 W/m²	Exposure limit in Russia (since 1970), Bulgaria Hungary
0.001 W/m²	"Precautionary Limit" in Austria, Salzburg City only
0.0009 W/m²	BUND 1997 (Germany) *Precaution recommendation only*
0.00001 W/m²	New South Wales, Australia

Table 1: International Exposure Standards and Guidelines

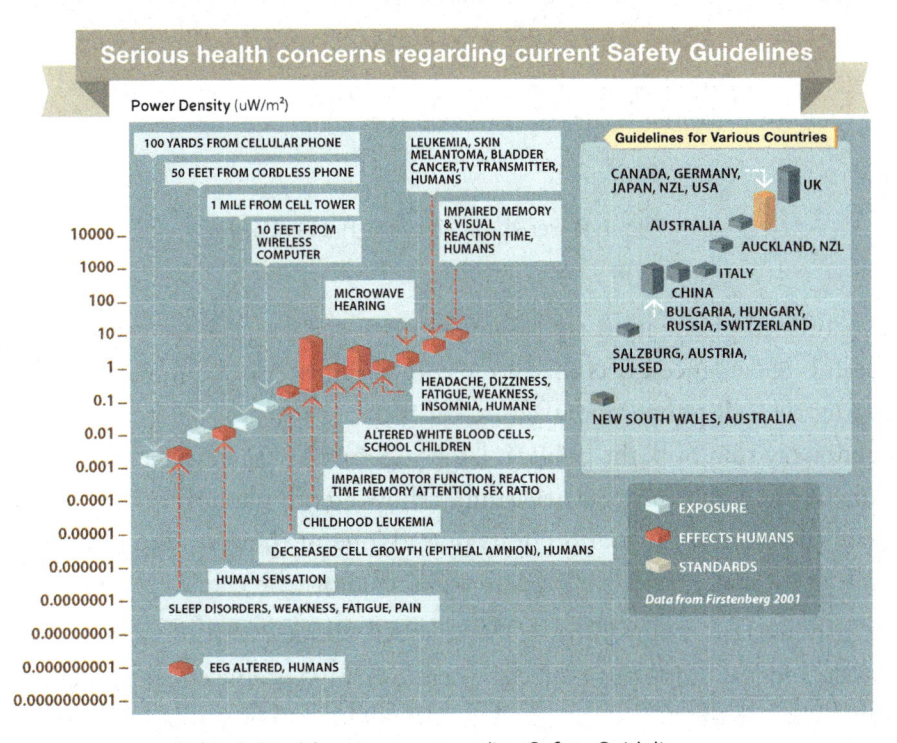

Table 2: Health concerns regarding Safety Guidelines

37

POWER ABSORBED BY HUMAN BODY

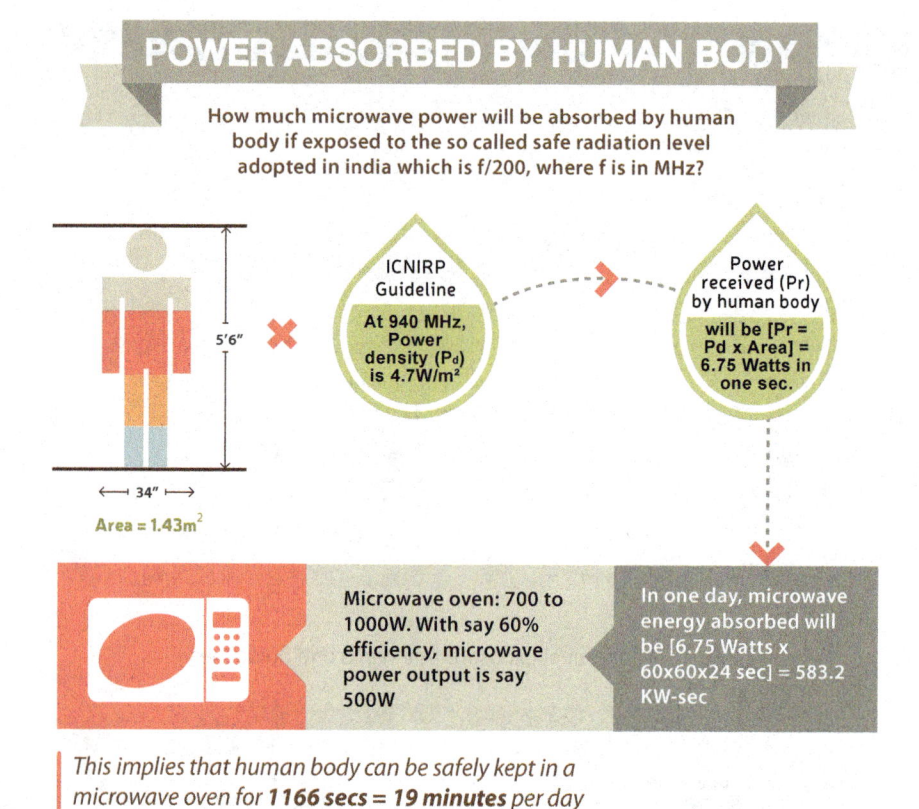

How much microwave power will be absorbed by human body if exposed to the so called safe radiation level adopted in india which is f/200, where f is in MHz?

5'6"
34"
Area = 1.43m²

ICNIRP Guideline
At 940 MHz, Power density (P_d) is 4.7W/m²

Power received (Pr) by human body
will be [Pr = Pd x Area] = 6.75 Watts in one sec.

Microwave oven: 700 to 1000W. With say 60% efficiency, microwave power output is say 500W

In one day, microwave energy absorbed will be [6.75 Watts x 60x60x24 sec] = 583.2 KW-sec

This implies that human body can be safely kept in a microwave oven for **1166 secs = 19 minutes** per day

Fig. 2: Power Absorbed by the Human Body

The cell phone industry is fully aware of the dangers. In fact, enough scientific evidence exists that some companies' service contracts prohibit suing the cell phone manufacturer or service provider, or joining a class action lawsuit. Still, the public is largely ignorant of the dangers, while the media regularly trumpets new studies showing cell phones are completely safe to use. Yet, Dr. Carlo points out, "None of those studies can prove safety, no matter how well they're conducted or who's conducting them."

There is no doubt in my mind that cell phone radiation is incredibly toxic and has cumulative effects. The US Army became concerned about the issue of the effects of EMR in the 1940's. They embarked upon a series of extensive investigations about of EMR on living organisms. In a lot of their experiments they did studies with the young of animals. The research conducted by the US Army demonstrated that microwaves created congenital deformities in young chick embryos.

So unless you assume personal responsibility for your health and do something to protect yourself you are regularly "microwaving" yourself and your loved ones and reducing your respective life spans, even if you do not die from serious illness. There are already significant number of successful lawsuits against mobile phone manufacturers with more to come.

REFERENCES

Abdel-Rassoul G, et al, Neurobehavioral effects among inhabitants around mobile phone base stations, Neurotoxicology, 28(2), 434-40, 2006

Agarwal Aet al Relationship between cell phone use and human fertility: an observational study, Oasis, The Online Abstract Submission System, 2006

Altamura G, Toscano S, Gentilucci G, Ammirati F, Castro A, Pandozi C, Santini M,Influence of digital and analogue cellular telephones on implanted pacemakers, European Heart Journal, 18(10), 1632-4161, 1997

Arthur Firstenberg, Microwaving Our Planet

Balmori, A. (2002). Evidence of a connection between sparrow decline and the introduction of phone mast GSM

Balmori A., Electromagnetic pollution from phone masts. Effects on wildlife, Pathophysiology 16 (2009) 191–199

Eger H., Hagen K. U., Lucas B., Vogel P., Voit H., The Influence of Being Physically Near to a Cell Phone Transmission Mast on the Incidence of Cancer, Published in Umwelt • Medizin • Gesellschaft 17,4 2004

Gandhi et al., IEEE Transactions on Microwave Theory and Techniques, 1996.

Hayes DL, Wang PJ, Reynolds DW, et al. Interference with cardiac pacemakers by cellular telephones. N Engl J Med. 1997 May 22;336(21):1473-9.

Hardell L, Carlberg M, Hansson Mild K. Pooled analysis of two case-control studies on the use of cellular and cordless telephones and the risk of benign brain tumours diagnosed during 1997-2003. Int J Oncol. 2006 Feb;28(2):509-18.

Hardell L, Mild KH, Carlberg M, Hallquist A. Cellular and cordless telephone use and the association with brain tumors in different age groups. Arch Environ Health. 2004 Mar;59(3):132-7.

Hardell L, Carlberg M, So¨derqvist F, Hansson Mild K, Morgan LL. Long-term use of cellular phones and brain tumours: increased risk associated with use for >/_10 years. Occup Environ Med 2007;64: 626e32.

Haumann Thomas, et al, " HF-Radiation levels of GSM cellular phone towers in residential areas"

Hutter HP et al, Tinnitus and mobile phone use, Occup Environ Med. 2010

Jokela K, Puranen L, Sihvonen A-P (2004) 'Assessment of the magnetic field exposure due to the battery current of digital mobile phones'. Health Physics 86: 56-66.

Lahkola A, Auvinen A, Raitanen J, et al. Mobile phone use and risk of glioma in 5 North European countries. Int J Cancer. 2007 Apr 15;120(8):1769-75.

Leigh Erin Connealy, MD EMFS: The Dangers of Modern Convenience

Lonn S, Ahlbom A, Hall P, Feychting M. Mobile phone use and the risk of acoustic neuroma. Epidemiology. 2004 Nov;15(6):653-9.

Mobile phone increases the temperature of the brain (Sciences et vie, No. 949 10/96) and raises blood pressure (publication of the University of Fribourg in the Lancet, 06/07 – 98).

Panda et al, Audiologic disturbances in long-term mobile phone users., J Otolaryngol Head Neck Surg., Chandigarh, 2010 Feb 1;39(1):5-11.

Salford, Leif G et al., Nerve Cell Damage in Mammalian Brain After Exposure to Microwaves from GSM Mobile Phones, Environmental Health Perspectives 111, 7, 881–883, 2003

Schreier N, Huss A, Roosli M. The prevalence of symptoms attributed to electromagnetic field exposure: a cross-sectional representative survey in Switzerland. Soz Praventivmed. 2006;51(4):202-9.

Schuz J, Jacobsen R, Olsen JH, Boice JD Jr, McLaughlin JK, Johansen C. Cellular telephone use and cancer risk: update of a nationwide Danish cohort. J Natl Cancer Inst. 2006 Dec 6;98(23):1707-13.

Salford, Leif G et al., Nerve Cell Damage in Mammalian Brain After Exposure to Microwaves from GSM Mobile Phones, Environmental Health Perspectives 111, 7, 881–883, 2003

Santini R, Santini P, Danze JM, Le Ruz P, Seigne M, Study of the health of people living in the vicinity of mobile phone base stations: Incidence according to distance and sex, Pathology Biology, 50(6), 369-73, 2002 27

Westerman R, Hocking B. Diseases of modern living: neurological changes associated with mobile phones and radiofrequency radiation in humans. Neurosci Lett. 2004 May 6;361(1-3):13-6.

Wilson BW, Stevens RG, Anderson LE eds (1990) Extremely low frequency electromagnetic fields: the question of cancer . Battelle Press, Columbus, Ohio

ABSTRACTS

Dr. Henry Lai, a well-known bioelectromagnetics researcher at the University of Washington, Seattle, has compiled a 97-page collection of abstracts from studies conducted between 1995 and 2000. The list, in pdf format, can be found on the Research page of the EMR Network's web site. As the web site points out, "80% of these studies demonstrate some kind of biological effect."

REPORTS

NIEHS (1998) Assessment of health effects from exposure to power-line frequency electric and magnetic fields. Portier CJ and Wolfe MS (eds) NIEHS Working Group Report, National Institute of Environmental

Health Sciences of the National Institute of Health, Research Triangle Park, NC, USA, pp 523.

ICNIRP (1998) International Commission on Non-Ionizing Radiation Protection

Guidelines for limiting exposure to time varying electric, magnetic and electromagnetic ields (up to 300 GHz). Health Physics 74(4), 494-522.

The Physiological and Environmental Effects of Non-Ionising Electromagnetic Radiation is a 34-page report issued in March 2001 by the European Parliament Directorate General for Research, Scientific and Technological Options Assessment (STOA). Written by Dr. Gerard Hyland, it pulls no punches in warning of the hazards of microwave radiation.

Potential and Actual Adverse Effects of Radiofrequency and Microwave Radiation at Levels Near and Below 2 uW/cm^2, is a 200-page report by Dr. Neil Cherry, of Lincoln University, New Zealand.

3. WHAT HARMFUL RADIATION BOMBARDS US DAILY?

Most of us remain ignorant of how much radiation we are exposed on a daily basis. In the account below, I have taken a sample of an average day in the life of someone who works in an office in Orchard Road in Singapore. All the words that are underlined and or in bold refer to sources of EMR and ionizing radiation. For simplicity I have chosen to underline sources of EMR and put in bold, sources of ionizing radiation.

YOUR DAILY DOSE OF RADIATION

You get up in the morning and put your trilling alarm clock to silent mode. Then you stretch wearily and get up. Little were you aware of how much EMR your little clock was pouring into your brain and how much it was contributing to your disturbed sleep. You rush to the toilet brush your teeth with an electric toothbrush, radiating your brain and mouth with more toxic EMR. After your shower you use your hair dryer to dry your hair, without recognizing that hairdryers emit a disproportionate amount of EMR toxicity.

You suddenly realize that you left your router on the whole night through. So you switch it off. That means all night long, radio waves were frying your brain and body. As you microwave your breakfast you further microwave your brain in the process because you are sitting next to the microwave oven at the table. You quickly scan through your mobile handphone (further increasing the dose of EMR exposure) and look at what messages are displayed.

You get in your car and turn it on. As you start travelling to work you pass through a few ERP gantries. You idly notice as you move along

43

Orchard Road, the cell phone towers that are sprouting up like metallic mushrooms. You feel the onset of a slight headache as you enter your office in the heart of Orchard. You shrug your headache off and attribute it to your disturbed sleep. Little did you know that the headache you were experiencing was due to electromagnetic sensitivity, as well as the restless sleep. They were both due to the EMR you were exposed to while you were sleeping (electric clock and wireless router).

You sit down at your computer in a room filled with four other people with computers and a wireless router. You are surrounded by radiation from the computers, the router and the cell phone towers. You suddenly remember that you have a dental appointment.

So you rush in to see the dentist, passing the ERP gantries and getting another major dose of microwave radiation. When your dentist examines you, she is concerned about one particular tooth. She wants to check on whether there is a hidden infection because you keep complaining to her about sensitivity in that region. She orders a dental X-ray (radiation). Fortunately there is nothing in the Xray plate to suggest an infection. You heave a sigh of relief. She does a routine dental scaling and gives you a special toothpaste to reduce the sensitivity.

You see a friend off at the airport that night. As you near the airport, you feel your residual headache coming back on. You body is responding to all the EMR coming from the radar systems and the wireless systems in the airport. In one day, you have been exposed to levels of radiation that an increasing amount of scientists are saying is toxic. According to the European Parliament, you have been exposed to radiation values of one million to one thousand million times higher than someone living and working in rural parts of Malaysia.

> **What you need to know**
>
> Sources of electric, magnetic fields that we are exposed to:
>
> Home: Electric and magnetic fields in homes depend on many factors, including the distance from local power lines, the number and type of electrical appliances in use in the home, and the configuration and position of household electrical wiring.

> Community: Electrical energy from generating stations is distributed to communities via high voltage transmission lines
>
> Workplace: Electric and magnetic fields exist around electrical equipment and wiring throughout industry.
>
> The following items also possess a significant electromagnetic field; television, electric razors, electric blankets, electric-power transmission and distribution networks, fluorescent lights, electric clocks, hair dryers electric heaters, microwave ovens and personal radio transmitters. Research appears to indicate that all of the above appliances at times have radiation levels that sometimes exceed current safety levels of radiation.
>
> Radiation Damage has been found involved in the following conditions:
>
> Miscarriage, birth defects, breast cancer (in both men and women), adult and childhood leukemia, hormone suppression, depression, suicide, Alzheimers disease, Parkinsons disease, Lou Gehrigs disease and ALS.
>
> DNA damage from radiation has been implicated in: Cancer, reduced fertility, permanent genetic damage
>
> **The Swedish government has stated that it will act on the assumption that a relationship exists between EMF exposure and childhood cancer, and a safety standard of 2 mG has been suggested.**
>
> International Commission on Non-Ionizing Radiation Protection has set arbitrarily so called International Standards and Safety Guidelines that are 1G to 5 G which is between 500 to 2500 times higher than the Swedish government standard (2 mG). If you followed their guidelines and thought you were safe you would be unpleasantly surprised.

THE ELECTROMAGNETIC SPECTRUM (SEE APPENDIX 2 FOR MORE INFORMATION)

Electromagnetic fields (EMF) consist of electric (E) and magnetic (H) waves travelling together. They travel at the speed of light and are characterised by a frequency and a wavelength. The frequency is simply the number of oscillations in the wave per unit time, measured in units of hertz (1 Hz = 1 cycle per second), and the wavelength is the distance travelled by the wave in one oscillation (or cycle).

EMFs are sometimes called radiation when the frequency is measured in kilohertz and above. EMFs are categorized, according to their frequency or wavelength, in the electromagnetic spectrum. The spectrum spans an enormous range of frequencies.

The lowest frequency EMFs (below 3000 Hz, or 3 kHz) are called extremely low frequency (ELF) fields. They are mainly generated by AC current devices and power lines and usually have a frequency of 60 Hz (North America) or 50 Hz (elsewhere).

Frequencies in the kHz (thousand hertz) and low MHz (megahertz, million hertz) region are called radio frequency (RF) fields or radiation and are used for radio and TV broadcasting and two-way radio systems.

The RF region is – arbitrarily – broken up into a further alphabet of frequencies like EHF, SHF, UHF, VHF, HF, MF, LF and VLF. In addition to radio transmitter equipment, also computer displays radiate RF waves in the kHz region in addition to the ELF fields associated with their AC power supply.

Frequencies in the high MHz and the GHz (gigahertz, billion hertz) region are called microwave (MW) fields or radiation and are used for cell phones, personal communication systems, microwave ovens and radar systems.

Frequencies above microwave and below visible light (10^{12} – 10^{14} Hz) are called infrared radiation. This type of EMF is radiant heat emitted from hot objects like ovens.

Visible light is a narrow band of frequencies around 10^{15} Hz. Visible light is emitted from atoms when electrons in their outer shells change orbits around the nucleus of the atom.

Frequencies in the spectral region above visible light are - according to increasing frequencies - ultraviolet, X-rays, gamma rays and nuclear radiation. These types of radiation are categorized as ionizing radiation, whereas all frequencies below ultraviolet are called non-ionizing radiation.

NATURAL VERSUS MAN MADE EMF IN THE NON-IONIZING SPECTRAL REGION

Few people realize how much and how abruptly we have changed the non-ionizing electromagnetic environment in less than a century. Actually, most of the changes have happened in the last few decades, and this development is accelerating at an exponential growth rate.

For billions of years, the electromagnetic environment was virtually "silent" in the spectral region below visible light which was the most abundant source of electromagnetic energy. Terrestrial biology has evolved in the long standing near-static magnetic field of the earth and the static electric field produced between the upper-atmospheric ionosphere and the earth.

Nature's sources of oscillating low frequency EMFs are few and extremely weak. The only sources are the sun, distant radio stars and other cosmic RF sources, and the terrestrial sources originating from lightning primarily in the tropics. Even the sun can not be considered a strong source of energy in the non-ionizing spectral region, making natural ambient levels so low that the possibilities of biological or even health effects are negligible.

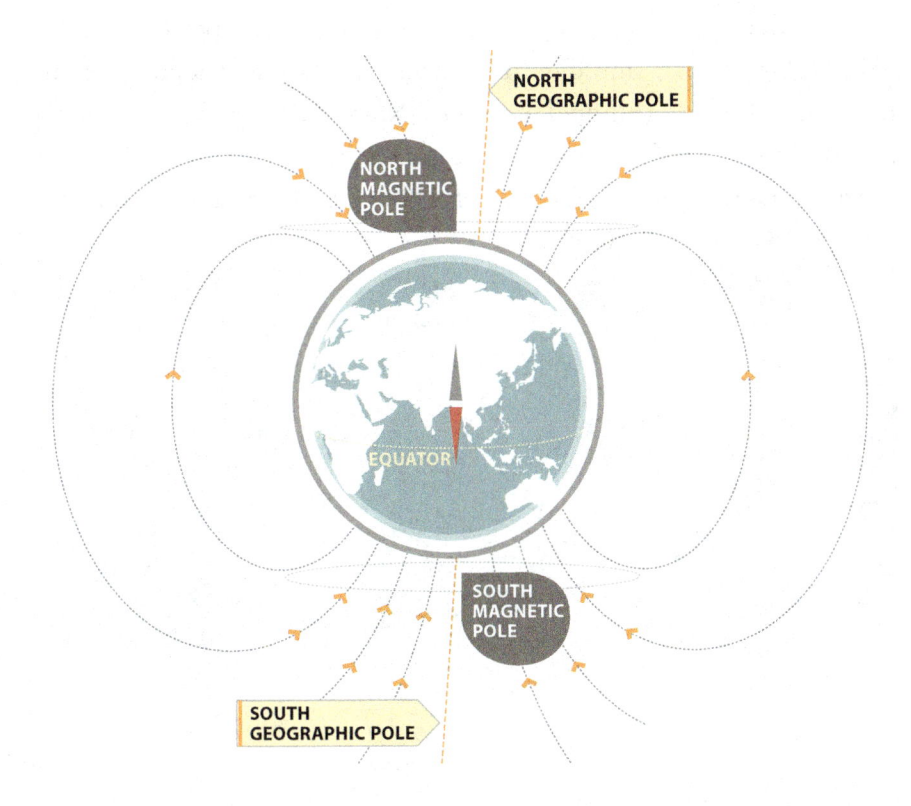

Fig.3: The earth's natural magnetic field

Grave concern was expressed by the European Parliament after a series of hearings on the matter of RF/MW frequency EMF health effects (Resolution B3-0280/92):

"Thus in the frequency range 100 kHz to 300 GHz, 50 years ago it was scarcely possible to measure 10 pW/cm^2 (10-12 W/cm^2) on the ground in our countries. Today, depending on the location, values one million to one thousand million times higher are recorded because of the explosion of telecommunications".

Perhaps even more important than the increase in EMF densities, most man made EMFs differ in one important aspect from natural EMFs. Man made sources radiate constant, regular oscillations or pulses of electromagnetic energy with distinct properties.

Natural sources of EMFs are oscillating at random with mixed, irregular frequencies and amplitudes. For example, AC powered devices emit highly regular, sinusoidal ELF fields at a constant amplitude and frequency of 50 Hz (Europe) or 60 Hz (North America). Digital mobile phones expose the heads of users to pulses of microwave, with carrier wave frequencies of around 900 or 1800 megahertz and modulation frequencies of 11, 22, 50 or 217 Hz.

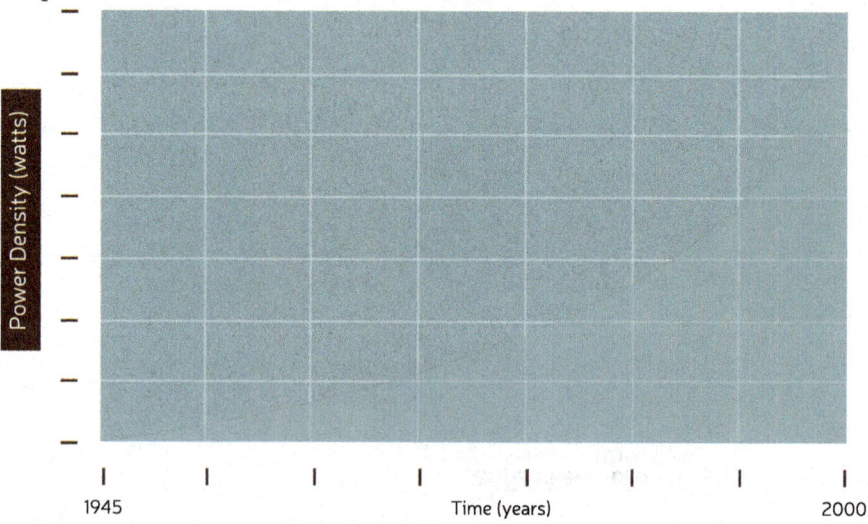

Fig. 4: Exponential growth in background non-ionizing radiation

The same applies to ELF fields produced by the dozens of electrical appliances owned by each household in the industrialized world and the power lines feeding those devices. Industrialized countries in North America, Western Europe and China emit so much ELF energy that it can be sensed by satellites in space.

SOURCES OF ELECTRICAL, MAGNETIC FIELDS

Naturally occurring 50/60 Hz electric and magnetic field levels are extremely low; of the order of 0.0001 V/m, and 0.00001 mT respectively. Human exposure to ELF fields is primarily associated with the generation, transmission and use of electrical energy. Sources and typical upper limits of ELF fields found in the community, home and work place are given below.

Community: Electrical energy from generating stations is distributed to communities via high voltage transmission lines. Electric and magnetic fields underneath overhead transmission lines may be as high as 12 kV/m and 30 µT respectively. Around generating stations and substations, electric fields up to 16 kV/m and magnetic fields up to 270 µT may be found.

Home: Electric and magnetic fields in homes depend on many factors, including the distance from local power lines, the number and type of electrical appliances in use in the home, and the configuration and position of household electrical wiring. Electric fields around most household appliances and equipment typically do not exceed 500 V/m and magnetic fields typically do not exceed 150 mT. In both cases, field levels may be substantially greater at small distances but they do decrease rapidly with distance.

Workplace: Electric and magnetic fields exist around electrical equipment and wiring throughout industry. Workers who maintain transmission and distribution lines may be exposed to very large electric and magnetic fields.

Within generating stations and substations electric fields in excess of 25 kV/m and magnetic fields in excess of 2 mT may be found. Welders can be subjected to magnetic field exposures as high as 130 mT. Near induction furnaces and industrial electrolytic cells magnetic fields can be as high as 50 mT. Office workers are exposed to very much smaller fields when using equipment such as photocopying machines and video display terminals.

ELECTROMAGNETIC RADIATION

As the computer visual display (VDU) unit became more common in the workplace, the issue of radiation hazards associated with the prolonged use of VDU's were tested by reputable laboratories and found to emit no detectable levels of X-rays.

A similar study by Canadian Radiation Protection Bureau researchers arrived at the same conclusion. World Health Organisation (WHO) experts endorsed similar findings. Given such reassurances, the temptation has been to conclude that VDU's are harmless. However, deeper more haunting statistics suggest that health problems from VDU's could arise from electromagnetic radiation.

The early research did not consider all the relevant data. Since 1979 small clusters of miscarriage and birth defects among VDU users in a dozen or more office locations have been reported. Due to the low level of X-ray radiation around VDU's, authorities often dismissed the increased incidence of these abnormalities as chance occurrences, while others argued alternately that the reported defects could be hereditary.

In 1982 Delagado and others reported powerful inhibitory effects on chicken embryos produced by weak 100Hz electromagnetic fields. The following year Ubeda and others also observed 'teratogenic" changes or monstrous mutations to chicken embryos exposed to low intensity pulsed electromagnetic fields of 100Hz.

The most deleterious effects were observed with a weak magnetic field strengths of about 1 micro Tesla, with stronger and weaker fields less effective. Since the original work of Delgado and co-workers, several more recent studies have confirmed that weak electromagnetic fields are capable of interacting with biological systems of specific frequencies and intensities.

Magnetic field strength pulses of up to 400,000 microtesla have been reported with VDU's. It follows that weak magnetic pulses will exist even at a considerable distance from the units. With approximately half the workforce using VDU's being women of childbearing age, the health implications are enormous.

McDonald and co-workers who studied births in the Montreal area in 1984, reported, that the rate of spontaneous abortion in 2609 current pregnancies with no VDU use was 5.7% compared to 8.3% for 588 with weekly exposure of less than 15 hours and 9.4% for 710 pregnancies with VDU use greater than 15 hours per week.

In 1988 Goldhaber and co-workers found in a case control study of pregnancy outcome that there exists: "Significantly elevated risk of miscarriage for working women who using VDU's for more than 20 hours per week during the first trimester of pregnancy compared to other working who reported not using VDUs". The increased risk could not be explained by age, education, occupation, smoking, alcohol consumption on other maternal characteristics.

The following items also possess a significant electromagnetic field; television, electric razors, electric blankets, electric-power transmission and distribution networks, fluorescent lights, electric clocks, hair dryers and electric heaters, microwave ovens, personal radio transmitters. Research done by Dr. Robert O. Becker M.D. appears to indicate that all of the above appliances at times have radiation levels that sometimes exceed current safety levels of radiation.

The rise of phenomena such as electromagnetic - hypersensitivity syndrome has been verified by several scientists, like Robert Becker M.D. and Dr. William Rae. It refers to the phenomena whereby an individual develops an allergy due to electro-magnetic fields. Chronic-fatigue syndrome has also been linked to electro-magnetic fields

> Brief exposure to microwaves, resulted in damage to nerve-cell structures that became visible, as a latent effect, only two to four months following the exposure. The nerve cell damage was visible in the brain, retina, optic nerve and cerebellum. It is interesting to note that clinical studies done by Dr. E. Courchesne of the Neuropsychology Research Laboratory at Children's Hospital Research Center, San Diego, report finding a specific pathological lesion in the cerebellum in fourteen of the eighteen autistic children be examined.

Dr. Hans-Anne Hanson of the Institute of Newsbiology at the University of Goteborg, Sweden began experimenting a newborn experimental animals:

> **Reported cases of foetal damage from VDU's:**
> 1. In 1979, four out of seven pregnant VDU operates who had worked on the classified advertising department of the Toronto Star gave birth to infants with defects. One had a club foot, another a deft palate, a third an underdeveloped eye and the fourth had multiple heart abnormalities. None of the mothers had smoked or taken drugs during the pregnancies. During that period, three other employees at the Star who didn't work on VDU's gave birth to normal babies.
> 2. Due to excessive fetal retardation and birth defects among the off-spring of women and animals exposed to radioactive fields in Eastern Europe, pregnant women in Czechoslovakia have been specifically prohibited from working in areas where the "safe" level of micro-waves was exceeded. **The Czech standard is one-thousandth the recommended American guideline.**

While Dr. Becker says that it is too premature to conclude that brain lesions of the autistic children match those of the experimental animals, he points out that the apparent onset of autism as a clinical condition in the 1940's coincided with the marked increase in the use of electromagnetic energy.

Electric power systems work at around 50hz, just above the naturally occurring frequency of 30hz generated by the Earth. Our bodies are tuned in to this low frequency so any frequency close to this is going to interfere with us. The 50hz Ac mains power system is used in the UK. ELF (Extremely Low Frequency) electromagnetic fields vibrating from 0.5 to 100hz, even if they are weaker than the Earth's field, interfere with the cues that keep our biological cycles properly timed. Chronic stress and impaired disease resistance results.

The frequencies that ELF AC systems use, are very close to the natural brain rhythms of cerebrate creatures such as man. The pineal gland is the principal structure in the brain that is directly sensitive to the earth's magnetic field. As a result it functions abnormally when exposed to abnormal fields close to its own. Since the pineal gland produces a host of psychoactive chemicals (such as melatonin, dopamine, serotonin and others) its abnormal functioning can cause neurological and behavioural problems.

In 1990 the United States Environmental Protection Agency stated "In conclusion, after an examination of the available data over the past 15 years, there is evidence of a positive association of exposure to magnetic fields with certain site-specific cancer, namely leukaemia, cancer of the central nervous system, and to a lesser extent, lymphomas. This is supported by many studies of children and adults across many different populations and sub-groups.

Electro-magnetic waves are known to cause changes in enzymes, hormones, and blood sugar levels, heating of the skin and its nerve endings. With more protracted exposure the coagulation of protein cells. These effects are amplified where a person is wearing metal objects, in the form of rings, zips, spectacles, hearing aids, buckles, watches, jewellery, or metal teeth fillings.

These metals produce secondary resonances within the bio-energetic field whenever one is in contact with anything electrical (AC or DC), or with electro magnetic radiation - for example household electrical equipment, cars, computers, fluorescent lighting, telephone switchboards; T.V.s, radios, V.D.U.s, microwave transmitters or ovens; and high voltage transformers, transmission cables and pylons.

The public is now understandably wary of safety assurances from "official" government scientific sources w.r.t. [with regard to] electromagnetic pollution. This skepticism is enhanced when views contrary to official perceived wisdom is [sic], at worst silenced or, at best, studiously ignored.

MARCH 2001 REPORT BY THE EUROPEAN PARLIAMENT STOA

Actually, we know what it does to us, so the results shouldn't come as any surprise.

The main problem isn't cancer, although the industry would like you to believe that. Thereafter they can pull out statistics showing how infrequently it occurs as a result of low-level radiation. Cancer takes a long time to develop.

Typically, other problems show up first: neurological, reproductive, and cardiac. Problems with severe headaches, sleep disturbances, memory loss, learning disabilities, attention deficit disorder, and infertility show up long before cancer. When cancer does appear, it's typically brain tumors, leukemia, and lymphoma.

For example, Kathleen Thurmond, M.D., in a 1999 talk, said,

A study presented by Dr. Ross Adey at the 1996 annual meeting of the Bioelectromagnetics Society in Victoria, B.C., Canada, showed a decrease in the incidence of brain tumors in rats chronically exposed to digital cellular telephone fields. However, there was no mention in his study of the increased incidence of spinal column tumors found in his research according to a reliable source. It would be standard scientific practice to at least note this finding regarding spinal column tumors. Dr. Ross Adey's research funding by Motorola has now been terminated.

Dr. Henry Lai was quoted in the London Times as saying, "They are asking me to change my whole interpretation of the findings in a way that would make them more favorable to the mobile phone industry. This is what happened in the tobacco industry. They had data in their hands but when it was not favorable they did not want to disclose it." The next chapter looks at what is a really safe dose of radiation for you and your family.

FIELD STRENGTH, SAFETY STANDARD AND OTHER SOURCES OF MAN MADE ELF FIELDS

Extremely low frequency (ELF; frequencies below 3 kHz) field strengths are measured in terms of the individual electric and magnetic components. Electric ELF fields are related to the voltage in conductors and are measured in volts per meter (V/m). Electric fields are present even if there is no current flowing. For example, an electric blanket will generate an electric field if plugged in, even if it is not turned on.

Magnetic fields are generated by the flow of current through conductors – the stronger the current the stronger the magnetic field. Magnetic field strengths are measured in the unit of milligauss (mG, 1/1000 of a gauss) since the gauss is a very big unit. Some scientists prefer another unit of measure called microtesla (mT, one millionth of a tesla); mG and mT relate to one another as follows: one mT equals 10 mG.

Most of the debate going on over acceptable thresholds of ELF fields is expressed in terms of magnetic field strengths. This is because the principal concern over the potential health effects relates to the magnetic component of the field. It's virtually impossible to shield, even with materials such as bricks, concrete, lead and earth. It penetrates deeply into cells, tissues, organs and the body of any person exposed.

This is why placing electrical transmission lines underground will not in itself reduce magnetic field exposures unless special engineering and design (shielding, phase cancellation, etc) are employed. The electric field, on the other hand, is much more effectively shielded; a house will shield out around 90 % of an electric field.

ELF fields envelop the surrounding space of a conductor or electrical device in the same way as the earth's magnetic field surrounds the planet. EMFs are not limited to power lines – they surround us in our work environment and homes, generated by hundreds of different appliances such as mobile phones and cordless phones, computers, laser printers, photocopiers, radar equipment, halogen and fluorescent lighting, hair dryers, vacuum cleaners, TV sets and stereo equipment, electronic games, electric shavers, electric blankets and heaters, food processors, coffee grinders, refrigerators, washers, tumble dryers, microwave ovens, and so on. ELF field strength can be measured with a gaussmeter, a fairly inexpensive device which can usually be purchased.

ELF field strengths drop off quickly with distance, and exposures to many appliances may be brief. However, research has indicated that even weak and short exposures have an effect, the effects seem to be cumulative, and multiple on/off exposures may induce stronger effects than continuous exposures.

These exposure standards were established in 1990, more or less arbitrarily. The problem with these standards is that numerous scientific studies have shown significant biological effects induced by EMF at field strengths thousands of times below these safety standards.

Epidemiologic studies have reported a substantially increased risk of cancer at much lower field strengths. In two Swedish residential studies published in 1993 and 1994, up to 3.8 times the expected rate of cancer was found in persons residing near power lines with time-weighted exposures of up to 4 mG. In a Danish residential study published in 1994, up to five times the expected rate of all cancers was found in people residing near power lines. In a Swedish occupational study, up to 5 times the expected rate of cancer was found in persons exposed to EMFs on the job. Table 1 shows the international standards and safety guidelines established by the International Commission on Non-Ionizing Radiation Protection.

INTERNATIONAL COMMISSION ON NON-IONIZING RADIATION PROTECTION		
EXPOSURE (50/60 HZ)	ELECTRIC FIELD	MAGNETIC FIELD
OCCUPATIONAL:		
WHOLE WORKING DAY	10 kV/m	5 G (5,000 mG)
SHORT TERM	30 kV./m	50 G (50,000 mG)
FOR LIMBS	--	250 G (250,000 mG)
GENERAL PUBLIC:		
UP TO 24 HOURS PER DAY	5 kV./m	1 G (1,000 mG)
FEW HOURS PER DAY	10 kV./m	10 G (10,000 mG)

*For electric fields of 10-30 KVM, field strength (kV/m) x hours of exposure should not exceed 80 for the whole working day. Whole body exposure to magnetic fields up to 2 hours per day should not exceed 50 G.

Table 3: ELF safety standards.

50/60-Hz Magnetic Fields for Some Appliances

485 Microwave Ovens Measured (outside 60-Hz fields, not inside the oven)		Maximum **Median** Minimum	263.9 mG 36.9 mG 0.9 mG	Maximum **Median** Minimum	17.2 mG 2.1 mG 0.2 mG
118 Analog Clocks Measured		Maximum **Median** Minimum	41.2 mG 14.8 mG 1.8 mG	Maximum **Median** Minimum	3.2 mG 0.3 mG 0.0 mG
383 Electric Ranges Measured		Maximum **Median** Minimum	28.6 mG 9.0 mG 0.5 mG	Maximum **Median** Minimum	6.2 mG 0.3 mG 0.0 mG
343 Color TVs Measureed		Maximum **Median** Minimum	18.6 mG 7.0 mG 0.4 mG	Maximum **Median** Minimum	1.4 mG 0.4 mG 0.02 mG
95 Digital Clock/ Clock-Radios Measured		Maximum **Median** Minimum	5.7 mG 1.3 mG 0.3 mG	Maximum **Median** Minimum	1.3 mG 0.2 mG 0.0 mG

Table 4: Typical ELF levels in mG (50/60 Hz) emitted by appliances and devices

On the weight of these studies, the Swedish government has stated that it will act on the assumption that a relationship exists between EMF exposure and childhood cancer, and a safety standard of 2 mG has been suggested.

To the average person, household and office appliances represent a more significant source of EMF exposure than power lines. As it appears from Table 2 above, many appliances emit EMR at intensities well above the 2 mG limit considered the defacto safety standard by Swedish authorities.

WHAT SHOULD BE DONE WHILE RESEARCH CONTINUES?

One of the objectives of the International EMF Project is to help national authorities weigh the benefits of using EMF technology against the detriment should any adverse health effects be demonstrated, and decide what protective measures, if any, may be needed. It will take some years for the required research to be completed, evaluated and published by WHO. In the meantime, WHO recommends:

Strict adherence to existing national or international safety standards: Such standards, based on current knowledge, are developed to protect everyone in the population.

Simple protective measures: Fences or barriers around strong ELF sources help preclude unauthorised access to areas where national or international exposure limits may be exceeded.

Consultation with local authorities and the public in siting new power lines: Obviously power lines must be sited to provide power to consumers. Despite the fact that ELF field levels around transmission and distribution lines are not considered a health risk, siting decisions are often required to take into account aesthetics and public sensibilities. Open communication and discussion between the electric power utility and the public during the planning stages can help create public understanding and greater acceptance of a new facility.

An effective system of health information and communication among scientists, governments, industry and the public can help raise general awareness of programmes to deal with exposure to ELF fields and reduce any mistrust and fears.

The human body is a physiological antenna. The pineal gland is also an antenna for the reception of thoughts. The cranium resonates in the microwave range! The human body is thus a detector of electromagnetic waves and it can react to EMR pollution in the same way that highly sensitive electronic circuitry can (such as a radio set).

The most intense and widespread EMR produced by man is in the 50 and 60 Hz range. These man-made fields are about 10 times stronger than those produced naturally. There is no place on Earth today that is not permeated with man-made EMR.

Our homes contain all kinds of devices that produces EMR that affects us cumulatively. Everything from toasters, to fridges, to ovens, to lamps, to unused wires in the walls, to TV sets produce EMR, even when they are off as long as they are plugged in!

An even higher source of EMR comes from power lines. Some of these produce such strong fields that they bend the Earth's ionosphere and can even cause satellites to dip in their orbits above them! Just think about those highways you may be traveling daily on.

They are usually lined with these transmission lines, and notice the microwave transmission towers! Individuals suffering from multiple allergies may be adversely affected by typical field strengths occurring in the vicinity of industrial radio frequency/microwave equipment and worksites.

The risk of increasing heart attack rates by increasing microwave exposure has been presented by Manfred Fritsch of the priv. Institut fur Baubiologische Anwendungen e.V., Stuttgarterstrasse 30, 70736 Fellbach Germany, in his 1994 survey monograph Gefahrdenherd Mikrowellen (ISBN 3-431-03345-8).

Karel Marha, while director of the Physical Hazards section of the Canadian Centre for Occupational Health and Safety, Hamilton Ontario Canada, presented the results of his earlier research.

The research demonstrated that digital, i.e., pulsed microwaves at low power could produce effects far in excess of the barely noticeable thermal effect predicted for a sine wave signal of equivalent power. The pulsed microwave was killing rats in a few seconds, evidently by inadvertent interference pattern "hot spots" of focused power burning holes in cell walls.

Health effects of stimulating increased microwave exposure by regulatory action must be assessed thoroughly before any change is approved.

GAUSS METERS: DETECTING NON-IONIZING ELECTROMAGNETIC RADIATION

Invisible electromagnetic radiation surrounds us each day, emanating from diverse sources such as power lines, home wiring, computers, televisions, microwave ovens, photocopy machines, and cell phones.

While undetectable to the eye, scientists have proposed that electromagnetic radiation may pose serious health effects, ranging from childhood leukemia to brain tumors. As scientists continue to unravel the precise health dangers of electromagnetic radiation, it makes good sense to avoid these potentially dangerous frequencies as much as possible. A gauss meter is a useful tool you can use to measure electromagnetic radiation in your home and work environments.

Using the gauss meter at varied locations, you can easily detect electromagnetic radiation "hot spots" where exposure to these ominous frequencies is the greatest. Armed with this crucial information, you can then avoid these areas, re-arranging furniture or electronic devices as needed in order to avoid unnecessary exposure to electromagnetic radiation.

USING A BATTERY OPERATED AM RADIO TO MEASURING NON-IONISING RADIATION

A simple way of measuring if you are outside the direct influence of the electric field of an appliance is outlined below; To check the level of radiation from your TV set, simply turn on your small battery-operated AM radio. Tune it to a spot on the dial where you cannot hear a station and turn the volume up to maximum.

Hold the radio about a foot away from the front of the TV and switch the TV on. If you then move the radio away from the TV you will reach a distance at which the noise disappears. This method can be used for devices that give off radio-frequency fields such as computers and stereos.

It will not work for devices that give off 60-H3 only such as electric stoves and hair dryers. The distance at which the "white" noise disppears is the safe distance. To be absolutely safe, pregnant mothers should minimize exposure to VDU's computers and micro-wave ovens

HOW ARE HUMANS AFFECTED BY IONIZING RADIATION?

The effects of a given dose of ionizing radiation on humans can be separated into two broad categories: Acute and Long-Term effects.

Although a dose of just 25 rems causes some detectable changes in blood, doses to near 100 rems usually have no immediate harmful effects. Doses above 100 rems cause the first signs of radiation sickness including:

- o nausea
- o vomiting
- o headache
- o some loss of white blood cells

Doses of 300 rems or more cause temporary hair loss. There is also more significant internal harm, including damage to nerve cells and the cells that line the digestive tract. Severe loss of white blood cells, which are the body's main defense against infection, makes radiation victims highly vulnerable to disease.

In time, for survivors, diseases such as leukemia (cancer of the blood), lung cancer, thyroid cancer, breast cancer, and cancers of other organs can appear due to the radiation received.

CHARACTERISTICS OF DNA DAMAGE BY RADIATION EXPOSURE

Deletion of DNA segments is the predominant form of radiation damage in cells that survive irradiation. It may be caused by

(1) misrepair of two separate double-strand breaks in a DNA molecule with joining of the two outer ends and loss of the fragment between the breaks or

(2) the process of cleaning (enzyme digestion of nucleotides--the component molecules of DNA) of the broken ends before rejoining to repair one double-strand break.

ACUTE EFFECTS OF RADIATION

The acute, or more immediately-seen effects of radiation can affect the performance of astronauts. These effects include skin-reddening, vomiting/nausea and dehydration. Other tissue and organ effects are possible. Another term: Acute Radiation Syndrome.

LONG TERM EFFECTS RADIATION

Given that only moderate doses of radiation are encountered (and thus acute effects are not seen) the long-term effects of radiation become the most important to consider. A more dangerous event may be the non-lethal change of DNA molecules which may lead to cell proliferation, a form of cancer. Research topic: The RBE of alpha particles on stem cells. These single and double strand breaks, or lesions, can be studied with the scanning tunneling microscope.

UNITS OF MEASUREMENT

The unit used to measure radiation dosage is the rem, which stands for roentgen equivalent in man. It represents the amount of radiation needed to produce a particular amount of damage to living tissue. The total dose of rems determines how much harm a person suffers.

At Hiroshima and Nagasaki, people received a dose of rems at the instant of the explosions, then more from the surroundings and, in limited areas, from fallout. Fallout is composed of radioactive particles that are carried into the upper atmosphere by a nuclear explosion and that eventually fall back to the earth's surface.

Is it known how much radiation a human can take? The biological effects of heavy particle ionizing radiation are approximately proportional to what is called Absorbed Dose (or simply dose). This is measured with instruments which detect the average energy deposited inside a small test volume.

The unit of dose is the gray (abbreviated Gy) which represents the absorption of an average of one joule of energy per kilogram of mass in the target material. This new unit has officially replaced the rad, an older unit (but still seen a lot in the radiation literature).

One gray equals 100 rads. Absorbed Dose was originally measured for x-rays and gamma radiation but has been extended to describe protons and HZE particles. When used in predicting biological damage, a further distinction must be made as to the "quality" of the radiation.

Although the Absorbed Dose of some radiation may be measured, another level of consideration must be made before the biological effects of this radiation can be predicted.

The problem is that although two different types of heavy charged particle may deposit the same average energy in a test sample, living cells and tissues do not necessarily respond in the same way to these two radiations. This distinction is made via the concept of Relative Biological Effectiveness (RBE) which is a measure of how damaging a given type of particle is when compared to an equivalent dose of x-rays.

RISK ESTIMATES OF LOW-LEVEL IONIZING RADIATION

There has never been any doubt about the disastrous short and long-term effects of exposure to high doses of ionizing radiation. The public acceptance of billion dollar investments in the fifties and sixties in a runaway build-up of nuclear weapons production, civilian nuclear energy, as well as nuclear medicine, was based on one premise.

Confident assurances of enthusiastic radiation experts had supported this premise. The premise was that added exposures at dose levels acceptable to industry would not be found detrimental to human health.

Are you concerned about the impact of cellular handphones on your brain? Have you wondered about the increasing incidence of brain tumours especially amongst the very young? The next chapter addresses this issue.

Already for these reasons one has to challenge the normally used procedure of the official commissions. Furthermore the radiation standards were derived from the Hiroshima/Nagasaki cohorts exposed to rather high doses (> 100 cGy). Occupational radiation exposure is normally at much lower doses (< a few cGy per year).

Thus for risk estimates at low doses one has to extrapolate from the observed cancer and mutation rates at high doses. Continuous and concurrent dosimetry for monitoring uranium miners, nuclear dockyard workers and workers in other nuclear facilities is far superior to retrospective dosimetry that is based on assumptions which are now in serious question.

Finally, good statistical practice says that you never extrapolate far beyond the range of the data when good data in the right range is available. This again suggests that the actual risks are many times the more than official ones. We are consistently exposed to levels of radiation and emf from 10 to 1000 times beyond safety levels.

REFERENCES

Adey WR: Frequency and Power Windowing in Tissue Interactions with Weak Electromagnetic Fields. Proceedings of the IEEE, 1980, vol. 63, no. 1, p. 119-125.

Adey WR: "Tissue interactions with non-ionizing electromagnetic fields", Physiol. Rev., 1981, vol. 61, p: 435.

Adey WR (1990) 'Electromagnetic fields, cell membrane amplification, and cancer promotion'. In: Wilson BW, Stevens RG, Anderson LE (eds) Extremely Low Frequency Electromagnetic Fields: the Question of Cancer . Battelle Press, Columbus, Ohio, pp 211-249

Balmori A., Electromagnetic pollution from phone masts. Effects on wildlife, Pathophysiology 16 (2009) 191–199

Cember, H. Introduction to Health Physics. New York: Pergamon Press, 1983.

Dan Bracken T. (1992): Experimental Macroscopic Dosimetry for Extremely-Low-Frequency Electric and Magnetic Fields, Bioelectromagnetics Supplement, v. 1, p. 15-26.

Hardell L, Mild KH, Carlberg M, Hallquist A. Cellular and cordless telephone use and the association with brain tumors in different age groups. Arch Environ Health. 2004 Mar;59(3):132-7.

Hardell L, Carlberg M, So¨derqvist F, Hansson Mild K, Morgan LL.

Long-term use of cellular phones and brain tumours: increased risk associated with use for >/_10 years. Occup Environ Med 2007;64: 626e32.

Haumann Thomas, et al, " HF-Radiation levels of GSM cellular phone towers in residential areas"

H. A. Pohl: Dielectrophoresis: The Behavior of Matter in Non-uniform Electric Fields, Cambridge University Press, 1978.

http://www.who.int/ionizing_radiation/about/what_is_ir/en/index.html

L. Thomas: The Youngest Science, Oxford University Press, Oxford, Melbourne, 1985, 276 p.

L. Wolpert: J. Theor. Biol., 1969, vol. 25, 1-47.

Leigh Erin Connealy, MD EMFS: The Dangers of Modern Convenience

Martin, A. and Harbison, S.A. An Introduction to Radiation Protection. 3rd ed. London: Chapman and Hall, 1986.

M. Grandolfo, S. M. Michaelson and A. Rindi, eds; Biological Effects and Dosimetry of Static and ELF Electromagnetic Fields, Plenum Press, New York, London, 1985.

R. Hölzel, I. Lamprecht: Electromagnetic fields around biological cells, Neural Network World, 1984, vol. 3, p. 327-337

Repacholi M and Greenebaum B (1998) Interaction of static and extremely low frequency electric and magnetic fields with living systems: health effects and research needs. Bioelectromagnetics (In press). (Summary report of WHO scientific review meeting on static and ELF held in Bologna, 1997

R. O. Becker, G. Selden: The Body Electric, Morrow, New York, 1985.

Schreier N, Huss A, Roosli M. The prevalence of symptoms attributed to electromagnetic field exposure: a cross-sectional representative survey in Switzerland. Soz Praventivmed. 2006;51(4):202-9.

Schuz J, Jacobsen R, Olsen JH, Boice JD Jr, McLaughlin JK, Johansen C. Cellular telephone use and cancer risk: update of a nationwide Danish cohort. J Natl Cancer Inst. 2006 Dec 6;98(23):1707-13.

Shapiro, J. Radiation Protection. Cambridge: Harvard University Press, 1972.

S. Rowlands: J. Biol. Phys., 1982, vol. 10, p. 199.

S. J. Webb: Phys. Rep., 1980, vol. 60, p. 201.

Smith C. W., Best S. (1990): Electromagnetic Man, Edited by J. M. Dent & Sons, Ltd. London, p.26.

ABSTRACTS

Dr. Henry Lai, a well-known bioelectromagnetics researcher at the University of Washington, Seattle, has compiled a 97-page collection of abstracts from studies conducted between 1995 and 2000. The list, in pdf format, can be found on the Research page of the EMR Network's web site. As the web site points out, "80% of these studies demonstrate some kind of biological effect."

REPORTS

The Physiological and Environmental Effects of Non-Ionising Electromagnetic Radiation is a 34-page report issued in March 2001 by the European Parliament Directorate General for Research, Scientific and Technological Options Assessment (STOA). Written by Dr. Gerard Hyland, it pulls no punches in warning of the hazards of microwave radiation.

4. HOW RADIATION DAMAGES OUR CELLS

THE POINT OF INTERACTION: THE CELL MEMBRANE

Man made EMFs differ in one important aspect from natural EMFs. They radiate with steady, regular oscillations or pulses with mainly constant frequencies and amplitudes. Natural EMFs are highly irregular with mixed, random frequencies, amplitudes and waveform.

Understanding how EMFs affect life processes requires a new paradigm in the understanding of life itself. Many scientists are still lost in inadequate classic beliefs of mechanistic chemistry. Chemistry does not recognize how fundamental electromagnetism is to biology and life. Throughout this century, most biologists and doctors have believed only chemical processes were involved in growth and healing. Our body consists of about 50,000 billion cells. Each of which is a living entity in itself. All cells are somehow interconnected, organized and specialized in a finely tuned equilibrium.

What you need to know

How radiation changes water molecules

Here is how the water molecules at the cellular membrane become chaotic;

This range of frequencies of RF phones is related with the ease of the movement of water dipoles resulting in chaotic Brownian movement of water molecules.

In the process of Brownian movement water molecules located in close proximity to each other develop the "friction effect."

This results in an increase of the level of absorption of EMF energy emitted by RF phone and in the generation of heat (called "dielectric loss.").

The water molecular arrangement becomes "chaotic".

> Here's what happens when your cell is exposed to cell phone radiation:
>
> A sensor in the membrane of the biological cell is triggered by the invading radio wave.
>
> The sensor is triggered by changes in the molecular nature of water that has become "chaotic".
>
> Once the membrane recognition occurs, a series of protective biochemical reactions are initiated inside the cell as a means of cellular protection.
>
> Among these protective reactions are stress protein responses that serve to effectively "harden" the cell membrane and disrupt active transport.
>
> The "membrane hardening" effect then causes an intracellular build-up of waste products and toxins, including highly reactive free radicals.
>
> These free radicals have been shown in studies to result in cellular dysfunction.

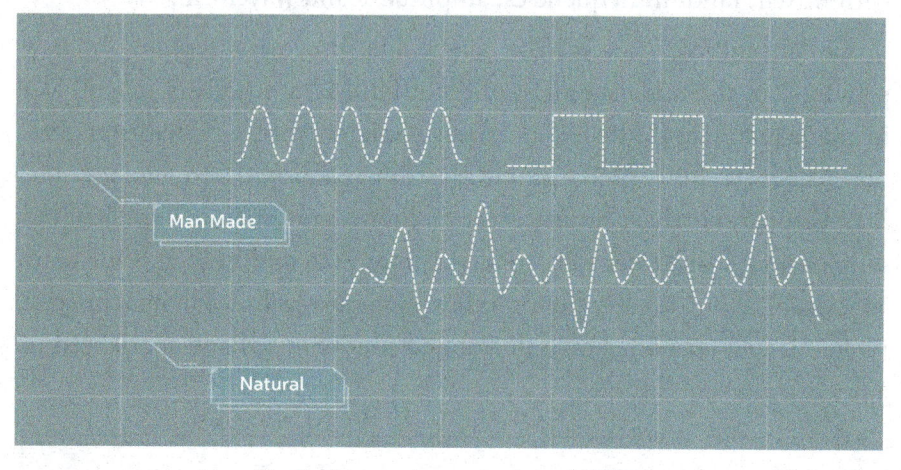

Fig. 5: Man-made versus natural EMFs.

These free radicals which are produced as a result of "membrane hatdening" have been shown in studies to result in cellular dysfunction (evidenced by studies showing leakage in the blood-brain barrier following EMR exposure). They also interfere with normal DNA repair processes (evidenced by studies showing the presence of micronuclei in cells following EMR exposure). These effects have been shown to lead to formation of disease. Several experiments have shown these effects eliminated when EMR exposure was removed.

HOW MAN-MADE RADIATION DAMAGES THE CELL-MEMBRANE

Mobile phones are held close to the body and are used frequently. These devices are potentially the most dangerous sources of electromagnetic radiation that the average person possesses. Regular GSM mobile phones and PDAs emit both pulsed radio waves (from the antenna) and ELF (from the battery circuits), and are especially dangerous. So how do these non-thermal effects of electromagnetic fields arise?

THE US ARMY STUDY

The research team at the Catholic University of America (CUA) in Washington, DC, was headed by Professor Litovitz. They had hypothesized and tested that an oscillating EMF needed to be steady for a certain minimum time period (approx. 1 second) for a biological response to occur.

This constancy allowed cells to discriminate external fields against thermal noise fields, even though the exogenous (external) field could be orders of magnitude lower than the endogenous noise field.

Also, the cell discriminates by spatial coherence, i.e., cooperative sensing by a large number of receptors in the cell membrane. Both temporal and spatial sensing is known from bacterial chemotaxis. The same medical community has largely ignored that EMFs have any influence on biology and health.

They have however for the past couple of decades been using steady pulsed ELF magnetic fields in hospital therapy to accelerate bone growth in fractures that otherwise would not heal. They have also used MW fields to accelerate healing of superficial wounds. The background for this treatment is easily explainable in light of recent laboratory experiments showing that ELF as well as MW frequency EMF accelerate cell growth.

The consequences relating to long term repeated, daily dosages of such cell growth accelerating stimuli is totally unpredictable. However we are dealing with something which might be potentially hazardous. Laboratory evidence has repeatedly shown that rapidly growing cells like transformed cancer cells are significantly more enhanced in growth rate than normal cells. This is among the factors adding to the conclusion that EMFs are possibly carcinogenic, as expressed by the NIEHS work group in July 1998.

Scientists are looking for the mechanism by which EMFs interact with living matter. They first recognize that living cells are bio-electro-chemical structures which interact with their environment in many ways including physically, chemically, biochemically and electrically.

Biological tissue consists of cells in a fixed structure, each cell surrounded by a conducting medium containing charges (ions) more or less free to move around each cell. A setup with extraordinary electrical properties which is ideal for establishing amplification effects with external constant oscillating EMFs.

The cell surface is a membrane equipped with thousands of fixed charges and receptors. It extends into the surrounding medium and senses the electrical, chemical and physical environment. It also provides feedback to the cell interior on any changes in this environment. The cell interior is filled with dispersed biological macromolecules with electrical properties, e.g., enzymes, proteins, DNA and RNA, which are responsive to electrical stimuli.

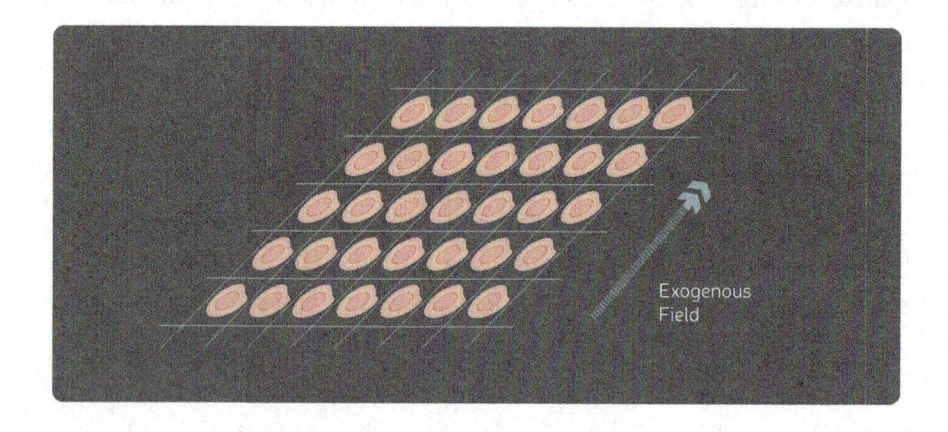

Fig. 6: Once we have many cells communicating in a living tissue, they may act cooperatively to amplify a faint electromagnetic sign

MESENCHYMAL MATRIX ,CELLULAR WATER AND RADIATION (APPENDIX 3)

The mesenchymal matrix is the substance between the cells. It is described as having sieve-like properties. It is a semi-fluid, liquid crystal medium, which occupies the space between the muscles, organs and cells of the body. Bio-energy signals traverse the matrix, and the term "matrix" is often used interchangeably with the term 'ground regulation system'. This structure is primarily filled with structured cellular water that is significantly affected by man-made radiation.

The proteoglycans, not only act as packaging material, they also account for the dielectric and other physiological properties of the matrix. For example, it is these properties, which enable bio-information transport, and selective excretion and resorbtion of urea and electrolytes in the kidneys.

Due to their polar nature and negative charges, proteoglycans bind with many water molecules to form the hydrated colloidal liquid crystalline gel that is responsible for the dielectric properties of the matrix.

GROUND REGULATION SYSTEM

The system of ground regulation has four main systems of communication. One is chemical, as in the many processes described above. The others are electrical impulses through the nerves, electrochemical synapses (found between fibroblasts and between functional organ cells helping them to act together), and electromagnetic vibrations. Thus the wide variety and vast quantity of internal and external information is coded and exchanged in only these four ways.

The following systems are intertwined with the Matrix;
1. Neurotransmitters and the Matrix
2. Electromagnetic fields and the Matrix -Man-made radiation can interfere with these naturally occurring electromagnetic fields by altering the arrangement of water molecules
3. Acupuncture points and the matrix – Both the water and chemical

contents of this bundle of blood vessels and nerves are electrically conductive. In comparison, fascia tissues are electrically resistant.

4. Biophotons
5. Neuropeptides and Neurotransmitters
6. Albert Szent-Gyorgi: Electronic Conduction - Water can form structures that transmit energy. These coherent water structures can also be disrupted by man-made radiation and initiate cellular damage.
7. Herbert Frohlich: Biological Coherence
8. Piezoelectric effect and the matrix – The combination of collagen fibers and sugar-protein complexes produces its highest piezoelectric energy values at 37° C, the temperature of the human body. And the liquid crystal molecular structure of water is highly ordered with minimum energy also at 37°.

PATHOLOGY ALTERS THE BIOMAGNETIC FIELD

In the 1920's and 1930's, a distinguished researcher at Yale University School of Medicine, Harold Saxon Burr, suggested that diseases could be detected in the energy field of the body before physical symptoms appear. Moreover, Burr was convinced that diseases could be prevented by altering the energy field. It also means that dramatically altering the electromagnetic field of the body can cause disease.

LATEST RESEARCH

Measurements of the biomagnetics of the heart and the brain led to a veritable explosion of research into biomagnetics. It turns out that biomagnetic fields are often more indicative of events taking place within the body than are electrical measurements at the skin surface . Every muscle produces magnetic pulses when it contracts. The larger muscles produce larger fields the smaller muscles, such as those that move and focus the eyes, produce tiny fields. This may be of interest to movement therapists, because we know that any movement from any part of the

body is broadcast into the space around the body as a "precise biomagnetic signature of movement."

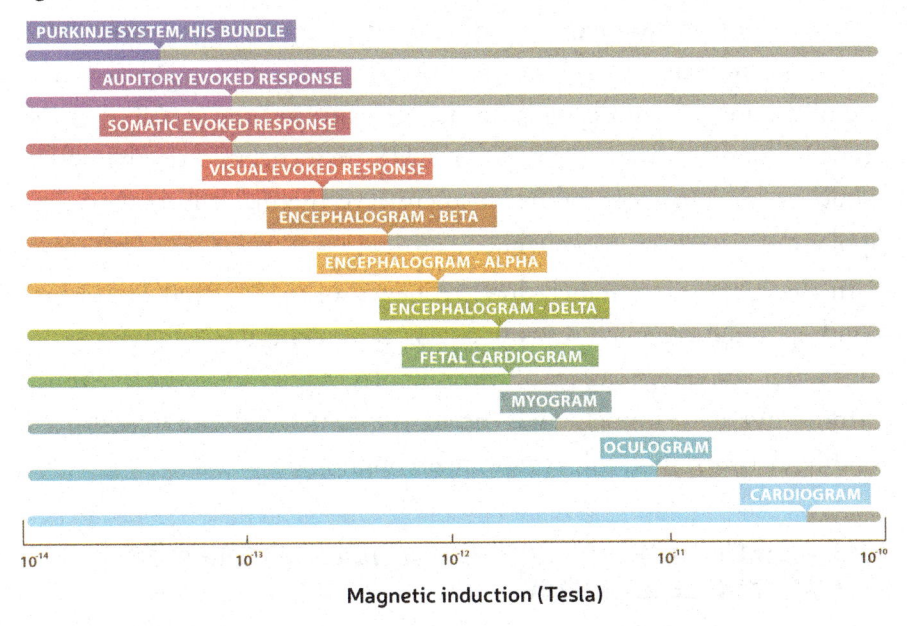

[Chart showing biomagnetic field strengths]

PURKINJE SYSTEM, HIS BUNDLE	
AUDITORY EVOKED RESPONSE	
SOMATIC EVOKED RESPONSE	
VISUAL EVOKED RESPONSE	
ENCEPHALOGRAM - BETA	
ENCEPHALOGRAM - ALPHA	
ENCEPHALOGRAM - DELTA	
FETAL CARDIOGRAM	
MYOGRAM	
OCULOGRAM	
CARDIOGRAM	

Magnetic induction (Tesla): 10^{-14} to 10^{-10}

Fig. 7: Relative strengths of the various biomagnetic fields measured in the spaces around the human body (Williamson & Kaufman, 1981).

Note that all the above fields are many times smaller then EMR emitted by Regular GSM mobile phones and PDAs emit both pulsed radio waves (from the antenna) and ELF (from the battery circuits). Is it any wonder than when the magnitude of the radio waves is thousand to a million times more then the naturally occurring fields of the human body, that the body goes into shock on exposure?

ELECTRODYNAMIC PROPERTIES OF WATER

There is scientific proof that extremely low frequency electromagnetic field can dramatically affect the dielectric permittivity and electrical conductivity of water and water based solutions.

Under the influence of applied EMF polar molecules tend to align themselves with the field. Although water has polar molecules, its hydrogen bonding network tends to oppose this alignment. The level to which a substance does this is called its dielectric permittivity.

Dependent on the frequency of applied EMF the dipole may move in time to the field, lag behind it or remain apparently unaffected. The ease of the movement depends on the viscosity and the mobility of the electron clouds. In the wide range of EMF frequencies lower than GHz frequency level (corresponding to microwave thermal effect) the water dipoles move in time to the field.

In the range of extremely low frequency of 0.1 – 1000 Hz (corresponding to the extremely low velocity of movement) the dynamic viscosity of water and the resistance of water dipoles to the alignment (dielectric permittivity) are extremely high (up to 108 at 0.1 Hz) due to hydrogen bonding between molecules (molecular coupling).

HOW RADIATION AFFECTS CELLULAR WATER CREATING FREE RADICALS AND CELL DAMAGE

In this extremely low range of frequency of applied EMF the water dipoles are able to move in time with low frequency electromagnetic field. As a result they can form multilayer molecular formations which oscillate in accordance with applied low frequency EMF.

In the higher frequency range of kHz to GHz generated by RF phones the reorientation process may be modeled using a "wait and switch" process where the water molecule has to wait until favorable orientation of neighboring molecules occurs and then the hydrogen bonds switch to new molecules. This range of frequencies of RF phones is related with the ease of the movement of water dipoles resulting in chaotic Brownian movement of water molecules.

In the process of Brownian movement water molecules located in close proximity to each other develop the "friction effect". That results in an increase of the level of absorption of EMF energy emitted by RF phone and in the generation of heat (called "dielectric loss."). This process creates free radicals in cellular tissue leading to radiation damage.

EMFS AND DISEASE

EMFs have been linked to a host of health concerns, including miscarriage, birth defects, breast cancer (in both men and women), adult and childhood leukemia, depression, suicide, Alzheimers disease, Parkinsons disease, Lou Gehrigs disease and ALS.

Increasingly researchers that are saying that, rather than causing direct harm, EMFs create subtle changes within the body that lead to serious diseases. Extensive research has been conducted examining the effects of EMFs on a hormone in the body called melatonin. Melatonin is secreted by the pineal gland in the center of the brain and controls the sleeping and waking cycles, enhances the immune system, lowers cholesterol and blood pressure and is a potent antioxidant that plays a part in preventing cancer, Alzheimers disease, Parkinsons disease, diabetes and heart disease.

DNA DAMAGE MAY CAUSE CANCER

There have been many studies suggesting that exposure to weak electromagnetic fields is associated with a small but significant increase in the risk of getting cancer (Wilson et al. 1990). This could be caused by gene mutations resulting from DNA damage. A gene is a section of DNA containing the information needed to make a particular protein or enzyme.

There is also a section that can turn the gene on or off in response to outside signals. The growth of an organism from a fertilised egg involves a hugely complex pattern of switching genes on and off that regulates growth, cell division and differentiation into specific tissues. DNA damage can sometimes give unregulated growth to form tumours. However, the effect may not be immediate. Cancer following exposure to chemical carcinogens such as asbestos may take many years to become rampant.

DNA DAMAGE REDUCES FERTILITY

The biological effects of electromagnetically induced DNA fragmentation may not be immediately obvious in the affected cells, since fragments of broken DNA can be rejoined and damaged chromosomes (elongated protein structures that carry the DNA) can be reconstituted.

However, there is no guarantee that they will be rejoined exactly as they were. Pieces may be left out (deletions) joined in backwards (inversions) swapped between different parts of the chromosome (translocations) or even attached to the wrong chromosome.

In most cases, the new arrangement will work for a while if most of the genes are still present and any metabolic deficiencies can often be made good by the surrounding cells. However, things go badly wrong when it comes to meiosis, which is the process that halves the number of chromosomes during the formation of eggs and sperm.

There is evidence from several independent studies in Australia, Hungary and the United States that this is already occurring. Heavy mobile phone use appears to reduce both the quantity and viability of sperm. According to the results for the most recent study by Dr Ashok Agarwal and coworkers at the Cleveland Lerner College of Medicine, they found that using a mobile phone for more than four hours a day was associated with a reduction in sperm viability and mobility of around 25%.

The statistical probability of these results being due to chance errors was one in a thousand. There is every reason to believe that human eggs may be similarly affected, but since they are formed in the embryo before the baby is born, the damage will be done during pregnancy but will not become apparent until the child reaches puberty.

THERE MAY ALSO BE PERMANENT GENETIC DAMAGE

Believe it or not, the electromagnetically induced loss of fertility is the good news since it means that badly damaged embryos are less likely to be conceived. The bad news is that any damaged genes needed for embryo development but not for normal egg or sperm function will not be weeded out in this way. They can still find their way into the foetus and cause permanent genetic damage.

The effect may not be apparent in the first generation since a non-functioning gene from one parent can often be offset if the other parent provides a good version of the same gene. In fact, serious trouble may not arise for many generations until by chance two faulty versions of the same gene end up in the same foetus. What happens then depends on the gene concerned, but it is unlikely to be beneficial and may be lethal. It is ironic that having only just discovered the human genome, we have already set about systematically destroying it.

ELECTROMAGNETIC EXPOSURE AND MOTOR ACCIDENTS

Only a small proportion of the population is electrosensitive in that they show obvious symptoms from electromagnetic exposure. However, everyone may affected without being aware of it, e.g. when using a mobile phone. According to the Royal Society for the Prevention of Accidents, you are four times more likely to have an accident if you use a mobile phone while driving.

This is not due to holding the phone since using a hands-free type makes no difference. It is also not due to the distraction of holding a conversation, since talking to a passenger does not have the same effect. This leads us to the conclusion that the electromagnetic radiation from the phone is the most likely culprit. This fits with the notion that spurious action potentials triggered by electromagnetic radiation creates a sort of 'mental fog' of false information that makes it harder for the brain to recognise weak but real stimuli.

HORMONE SUPPRESSION

Studies indicate that EMF exposure can shut down melatonin secretion in the body. Researchers at the University of North Carolina believe that the decreased levels of melatonin stemming from EMF exposure may cause depression and suicide.

Researchers compared levels of EMF exposure and rate of suicide among more than 5,000 electrical workers and an equal number of non-electrical workers. They found that the suicide rate of the electrical workers was twice that of the control group.

Robert P. Liburdy of Lawrence Berkley National Laboratory found that exposure to 12mG EMFs can suppress the ability of both melatonin and the hormone-emulating drug tamoxifen to shut down the growth of cancer cells.

The following chapter looks at how radiation impacts the ionic calcium distribution. This is the mechanism through which rather than causing direct harm, EMFs create subtle changes within the body that lead to serious diseases. The mechanisms by which the immune system is damaged, cholesterol and blood pressure increased and how cancer, Alzheimers disease, Parkinsons disease, diabetes and heart disease develop are examined in the next chapter.

REFERENCES

Adey WR: Frequency and Power Windowing in Tissue Interactions with Weak Electromagnetic Fields. Proceedings of the IEEE, 1980, vol. 63, no. 1, p. 119-125.

Adey WR: "Tissue interactions with non-ionizing electromagnetic fields", Physiol. Rev., 1981, vol. 61, p: 435.

Adey WR (1990) 'Electromagnetic fields, cell membrane amplification, and cancer promotion'. In: Wilson BW, Stevens RG, Anderson LE (eds) Extremely Low Frequency Electromagnetic Fields: the Question of Cancer . Battelle Press, Columbus, Ohio, pp 211-249

Alberts et al. (2002) Molecular Biology of the Cell . (Garland Science, New York)

A. F. Lawrence, W. R. Adey: Nonlinear Wave Mechanisms in Interactions between Excitable Tissue and Electromagnetic Fields, Neurological Research, v. 4, n. 1/2, 1982, p. 115-153.

A. Gurvich: Selected Works, Meditsina, Moscow, 1977.

A.L. Thomasset, Lyon Médical, Vol. 21, 1962; R.O.Becker and D.G. Murray, Transactions of the New York Academy of Sciences, Vol. 29, 1967;

A. Pischinger: Matrix and Matrix Regulation, Haug, Brussels, 1991.

Bary W. Wilson, Richard G. Stevens, Larry E. Anderson, editors:

Extremely Low Frequency Electromagnetic Fields: The Question of Cancer. Batelle Press, 505 King Avenue, Columbus Ohio, 43201-2693 tel. 1-800-451-3543 or 614-424-5263.

Blank M, Goodman R, Electromagnetic fields stress living cells, Pathophysiology 16 (2009) 71–78,

Diem E, Schwarz C, Aldkofer F, Jahn O, Rudiger H (2005) 'Non-thermal DNA breakage by mobile phone radiation (1800 MHz) in human fibroblasts and in transformed GFSH-R17 rat granulosa cells in vitro'. Mutation Research/Genetic Toxicology and Environmental Mutagenesis 583: 178-183

Elenkov IJ, Iezzoni DG, Daly A, Harris AG, Chrousos GP. "Cytokine dysregulation, inflammation and well-being". Neuroimmunomodulation. 2005;12(5):255-69

F.-A. Popp, K. H. Li and Q. Gu eds., Recent Advances in Biophoton Research and its Applications, , World Scientific, Singapore, 1992, 504 p.

Gallagher, J.T., Lyon, M. (2000). "Molecular structure of Heparan Sulfate and interactions with growth factors and morphogens", in Iozzo, M, V.:

Proteoglycans: structure, biology and molecular interactions. Marcel Dekker Inc. New York, New York, 27-59.

Ha B-Y (2001) 'Stabilization and destabilization of cell membranes by multivalent ions'. Phys. Rev . E. 64: 051902 (5 pages)

H. A. Pohl: Natural oscillating fields of cells, In: Coherent excitations in Biological Systems. (H.Frölich, F. Kremer, eds.), Springer Verlag, Berlin, 1983, 199-210.

H. A. Pohl: Natural ac electric fields in and about cells, Phenomena, v. 5, 1983, p. 87-103.

H. A. Pohl: Natural electrical RF oscillation from cells, J. Bioenerg. Biomembr., 1981, vol. 13, p.149-169.

H. Selye: The Stress of Life, McGraw-Hill Book Company, New York, Toronto, London, 1956.

H. Heine, P. Anastasiadis eds, Gustav Fischer, Stuttgart, Jena, "Normal Matrix and Pathological Conditions", New York, 1992.

Iozzo, R. V. (1998). "Matrix proteoglycans: from molecular design to cellular function". Annu. Rev. Biochem. 67: 609-652.

J. Aschoff: Comparative Physiology: Diurnal rhythms, Ann. Rev. Physiol., 1963, vol. 25, p. 581.

Karsenty G, Park RW (1995). "Regulation of type I collagen genes expression". Int. Rev. Immunol. 12 (2-4): 177–85.

Liotta LA, Tryggvason K, Garbisa S, Hart I, Foltz CM, Shafie S (1980). "Metastatic potential correlates with enzymatic degradation of basement membrane collagen". Nature 284 (5751): 67–8.

Liboff AR, McLeod BR, Smith SD (1990) 'Ion cyclotron resonance effects of ELF fields in biological systems'. In: Wilson BW, Stevens RG, Anderson LE (eds) Extremely Low Frequency Electromagnetic Fields: the Question of Cancer . Battelle Press, Columbus, Ohio, pp 251-289

Lin H, Blank M, Rossol-Haseroth K, Goodman R (2001) 'Regulating genes with electromagnetic response elements'. J Cellular Biochem 81: 143-148

Lodish H, Berk A, Matsudaira P, Kaiser CA, Krieger M, Scott MP, Zipursky SL, Darnell J. Molecular Cell Biology, 5th, New York: WH Freeman and Company, 197–234.

O'Brien SM, Scott LV, Dinan TG. "Cytokines: abnormalities in major depression and implications for pharmacological treatment". Hum Psychopharmacol. 2004 Aug;19(6):397-403.

Peach et al 1993. Identification of hyaluronic acid binding sites in the extracellular domain of CD44. The Journal of Cell Biology, Vol 122, 257-264

Pert CB, Ruff MR, Weber RJ, Herkenham M. Neuropeptides and their receptors: a psychosomatic network. J Immunol. 1985 Aug;135(2 Suppl):820s-826s

Ruff M, Schiffmann E, Terranova V, Pert CB.Neuropeptides are chemoattractants for human tumor cells and monocytes: a possible mechanism for metastasis. Clin Immunol Immunopathol. 1985 Dec;37(3):387-96

P.J. Rosch and M.S. Markov (eds), Bioelectromagnetic Medicine, Marcel Dekker, New York, NY (2004);

Panagopoulos DJ, Chavdoula ED, Nezis IP, Margaritis LH (2007) 'Cell death induced by GSM 900-MHz and DCS 1800-MHz mobile telephony radiation'. Mutation Research 626: 69-78

Plopper G (2007). The extracellular matrix and cell adhesion, in Cells (eds Lewin B, Cassimeris L, Lingappa V, Plopper G). Sudbury, MA: Jones and Bartlett.

R. Hölzel, I. Lamprecht: Electromagnetic fields around biological cells, Neural Network World, 1984, vol. 3, p. 327-337

Repacholi M and Greenebaum B (1998) Interaction of static and extremely low frequency electric and magnetic fields with living systems: health effects and research needs. Bioelectromagnetics (In press). (Summary report of WHO scientific review meeting on static and ELF held in Bologna, 1997

Sage C, Johansson O, Sage SA (2007) 'Personal digital assistant (PDA) cell phone units produce elevated extremely low frequency electromagnetic field emissions'. Bioelectromagnetics . DOI 10.1002/bem.20315 Published online in Wiley InterScience (www.interscience.wiley.com)

Smith SD. McLeod BR, Liboff AR (1993) 'Effects of SR tuning 60Hz magnetic fields on sprouting and early growth of Raphanus sativus'. Bioelectrochem Bioenerg 32: 67-76

W. Kroy: The Use of Optical Radiation for Stimulation Therapy, In: Electromagnetic Bio-Information (F.-A. Popp et al. ed.), Urban & Schwarzenberg, Munchen, 1989, p. 200-212.

W. Reich: The Cancer Biopathy, Orgone Institute Press, New York, 1948.

Westerman R, Hocking B. Diseases of modern living: neurological changes associated with mobile phones and radiofrequency radiation in humans. Neurosci Lett. 2004 May 6;361(1-3):13-6.

Wilson BW, Stevens RG, Anderson LE eds (1990) Extremely low frequency electromagnetic fields: the question of cancer . Battelle Press, Columbus, Ohio

5. CALCIUM AND RADIATION TOXICITY

So in my practice, I have seen many individuals who were affected by the negative effects of EMR. I started doing a literature review to determine the best possible treatment protocols. Based on the above, I developed a protocol to screen my clients routinely for radiation toxicity. I found that about 35% of those with chronic and terminal illness had radiation poisoning. This was from a database of 7, 000 individuals who were screened.

> **What you need to know**
>
> The universal chemical link between EMFs and life processes is believed by many scientists to be ions, especially calcium ions. External EMFs clearly affect the electrical properties and ion distribution around cells. The strange non-thermal biological effects of electromagnetic fields have puzzled scientists for decades and, until now, there has been no clear explanation. Well-replicated studies have shown that weak electromagnetic fields remove calcium ions bound to the membranes of living cells. Outlined below are functions of calcium ions.
>
> - Calcium functions as a ubiquitous intracellular messenger in a wide variety of cellular responses, including secretion and cell proliferation.
> - In nerve cells, calcium influx has been shown to be involved in the initiation of neurotransmitter secretion.
> - Many enzymatic processes are regulated by calcium. Calcium has been shown to modify gene transcription.
> - Calcium ions in partnerships with biomolecules have been shown to control the proliferation of non-tumorigenic cells in vitro and in vivo.
> - Calcium ions are involved in cell homeostasis which controls a variety of cellular responses determining the health of the cell.
> - It is widely accepted that calcium plays a central role in the development of the immune system response.

> Numerous scientific studies have demonstrated that EMFs can alter the membrane ion pumps. Ion pumps are responsible for pumping calcium, sodium and potassium in and out of the cells. Effects have been shown at low current densities, thousands of times lower than currents induced by MW fields. ELF fields have been shown to have the same effect.

The information about the calcium ions below made sense to me about why the supplements proved to be useful. In Chapter 8, we will look at supplements that have proven to be useful against radiation toxicity.

> **Case History 1**
>
> Fifteen years ago, I had the interesting experience of working with a mum to be who was a newscaster. She worked in a major newsgroup room with many other female colleagues. She mentioned that all her female colleagues who worked with her, had babies with congenital birth defects. She wanted a diet and nutrition program so that her baby would be normal. She came in about six months before her intended time to get pregnant. I put her on a tailored diet which was more macrobiotic and a tailored program supplements to support her system. Her baby when she delivered was normal. She was delighted.

> **Case History 2**
>
> Very recently my wife developed severe mouth ulcers that were not improving. As a result she could not eat and she was losing weight for 2 weeks. I finally tested her and confirmed that she had radiation toxicity. We worked out a specific treatment protocol that involved cellular hydration, anti-oxidants and nutrition and in 4 days her system kicked in and her immune function improved. These supplements are outlined in Chapter 7

THE CHEMICAL LINK: CALCIUM IONS AND ION DISTRIBUTION ACROSS THE CELL MEMBRANE

The universal chemical link between EMFs and life processes is believed by many scientists to be ions, especially calcium ions. External EMFs clearly affect the electrical properties and ion distribution around cells. Virtually all physiological processes in our body involve ions.

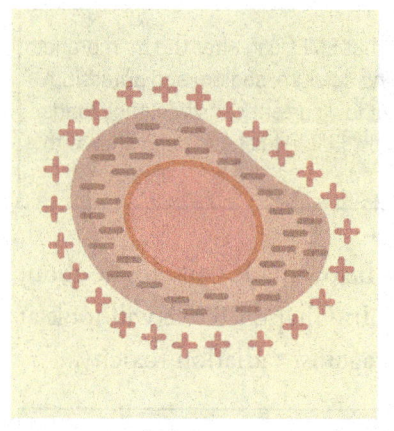

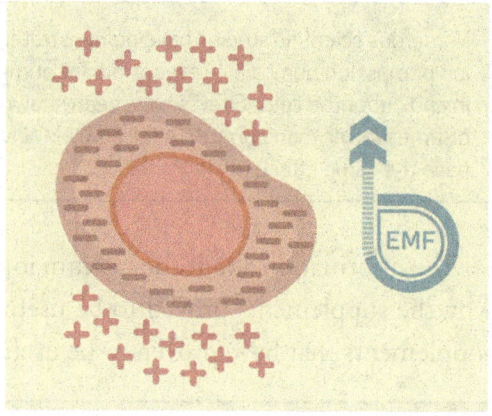

Fig. 8a: A normal resting cell: a cell with a uniform distribution of charges surrounding the negatively (−) charged membrane. (left)

Fig. 8b: A cell influenced by EMFs: a cell both with negative charges in the membrane and positive charges concentrating in the direction of the exogenous field.(right)

One of the most experienced researchers in the field, Dr. Ross Adey, in 1988 presented a three step model involving calcium ions. This model could explain observed EMF induced biological effects. Key to the model is the activation of intracellular messenger systems (adenylate cyclase and protein kinase) by calcium in a stimulus amplification process across the cell membrane. Numerous scientific studies have demonstrated the physiological importance of calcium.

Calcium functions as a ubiquitous intracellular messenger. For example, in 1947 it was shown that an intracellular injection of a small amount of calcium causes a skeletal muscle to contract. In recent years it has become clear that calcium acts as an intracellular messenger in a wide variety of cellular responses, including secretion and cell proliferation.

In nerve cells, calcium influx has been shown to be involved in the initiation of neurotransmitter secretion. The calcium enters the cells through voltage gated ion channels that open when the plasma membrane of the nerve terminal is depolarized by an invading action potential. Another function of calcium in all cells is to regulate metabolic processes in conjunction with the calcium-binding protein calmodulin.

Many enzymatic processes are regulated by calcium. Calcium has been shown to modify gene transcription. Thus, induced alteration of intracellular calcium concentrations which disrupt the homeostasis of the cell, has serious consequences for the health and future development of the cell.

Calcium ions in partnerships with biomolecules have been shown to control the proliferation of non-tumorigenic cells in vitro and in vivo. The evidence points to calcium and a biomolecule called AMP being co-generators of the signal committing the cell to DNA synthesis. Calcium influx in a cell stimulates proliferation, whereas calcium efflux does the opposite.

Researchers linked intracellular calcium levels to the future of damaged cells between becoming transformed (cancer) or dying by apoptosis (the healthy situation). Another study described the importance of calcium ions for cell homeostasis which controls a variety of cellular responses determining the health of the cell. Hence, reductions in intracellular calcium have a very important effect.

Calcium ions are involved in the function of gap junctions or protein structures which link adjacent cells and provide a channel for the passing of messenger molecules. The gap junction can open and close to control the flow.

The opening and closing is regulated by calcium ion concentration. Thus, calcium plays another key role in maintaining or interrupting the communication mechanisms for maintaining the health of cells. Gap junctions are used to sense differences between cells and to initiate corrections in regulatory behavior as necessary.

It is widely accepted that calcium plays a central role in the development of the immune system response. An elevation of calcium ions is a nearly universal feature associated with activation of cells of the immune system.

Using T-cell human leukemia cells, Lindstrom et al. (1995) replicated and extended the research of other scientists and showed, that oscillating low level EMFs produce the same calcium ion reaction as does an antibody.

Numerous scientific studies have demonstrated that EMFs can alter the membrane ion pumps. Ion pumps are responsible for pumping calcium, sodium and potassium in and out of the cells.

Effects have been shown at low current densities, thousands of times lower than currents induced by MW fields. ELF fields have been shown to have the same effect. Reference is made to a scientific paper from 1992 mentioning 10 different laboratories which have demonstrated these effects of calcium.

OVERVIEW OF EMF DAMAGE MODEL

Well-replicated studies have shown that weak electromagnetic fields remove calcium ions bound to the membranes of living cells. This will make the membranes more likely to tear, develop temporary pores and leak.

DNAase (an enzyme that destroys DNA) leaking through the membranes of lysosomes (small bodies in living cells packed with digestive enzymes) explains the fragmentation of DNA seen in cells exposed to mobile phone signals. When this occurs in the germ line (the cells that give rise to eggs and sperm), it reduces fertility and predicts genetic damage in future generations.

Leakage of calcium ions into the cytosol (the main part of the cell) acts as a metabolic stimulant. This accounts for reported accelerations of growth and healing. It also however promotes the growth of tumours. Leakage of calcium ions into neurones (brain cells) generates spurious action potentials (nerve impulses) accounting for pain and other neurological symptoms in electro-sensitive individuals.

It also degrades the signal to noise ratio of the brain making it less likely to respond adequately to weak stimuli. This may be partially responsible for the increased accident rate of drivers using mobile phones.

A more detailed examination of the molecular mechanisms explains many of the seemingly weird characteristics of electromagnetic exposure. These include why weak fields are more effective than strong ones, why some frequencies such as 16Hz are especially potent and why pulsed fields do more damage.

The strange non-thermal biological effects of electromagnetic fields have puzzled scientists for decades and, until now, there has been no clear explanation. In this we will look at a theory proposed by Goldsworthy (2006) that looks at how virtually all of these effects arise. Firstly, it is not only humans that are affected.

Well-researched responses in other organisms include the more rapid growth of higher plants (Smith et al. 1993; Muraji et al. 1998; Stenz et al. 1998), yeast (Mehedintu and Berg 1997) and changes in the locomotion of diatoms (McLeod et al. 1987). The last two are significant because they are both single cells, implying that the effects occur at the cellular level. Furthermore, we can explain virtually all of the electromagnetic effects on humans in terms of changes occurring at the cellular level that may then affect the whole body.

WEAK ELECTROMAGNETIC FIELDS RELEASE CALCIUM FROM CELL MEMBRANES

The first clue came from Suzanne Bawin, Leonard Kaczmarek and Ross Adey (Bawin et al. 1975), at the University of California. They found that exposing brain tissue to weak VHF radio signals modulated at 16Hz (16 cycles per second) released calcium ions (electrically charged calcium atoms) bound to the surfaces of its cells.

Carl Blackman at the U.S. Environmental Protection Agency in North Carolina followed this up with a whole series of experiments testing different field-strengths and frequencies (Blackman et al. 1982). Blackman came to the surprising conclusion that weak fields were often more effective than strong ones. The mechanism was unknown at the time and it was thought to be a trivial scientific curiosity, but as we will see, it has huge significance for us all.

THE LOSS OF CALCIUM MAKES CELL MEMBRANES LEAK

Calcium ions bound to the surfaces of cell membranes are important in maintaining their stability. They help hold together the phospholipid molecules that are an essential part of their make-up. Without these ions, cell membranes are weakened and are more likely to tear under the stresses and strains imposed by the moving cell contents (these membranes are only two molecules thick!). The resulting holes are normally self-healing they still increase leakage while they are open. This phenomenon can explain the bulk of the known biological effects of weak electromagnetic fields.

MEMBRANE LEAKAGE DAMAGES DNA

Leaks in the membranes surrounding lysosomes (tiny particles in living cells that recycle waste) can release digestive enzymes, including DNAase (an enzyme that destroys DNA). This explains the serious damage done to the DNA in cells by mobile phone signals.

Panagopoulos et al. (2007) exposed adult Drosophila doc 4 melanogaster (an insect widely used in genetic experiments) to a mobile phone signal for just six minutes a day for six day This process broke into fragments the DNA in the cells that give rise to their eggs and half of the eggs died.

Diem et al. (2005) also found significant DNA fragmentation after exposing cultured rat and human cells for 16 hours to a simulated mobile phone signal. Research shows that exposing human cells for 24 hours to simulated mobile phone signals gave DNA fragmentation similar to that due to the gamma rays from a radioactive isotope! (Gamma rays also make lysosome membranes leak.)

EFFECTS ON METABOLISM

Another major effect of electromagnetic radiation is the leakage of free calcium ions, either through the cells' external membranes or those surrounding internal 'calcium stores'. This can have dramatic effects on many aspects of metabolism.

It also explains most of the mysterious but well-documented physiological effects of electromagnetic fields. These include stimulations of growth, an increased risk of cancer and symptoms suffered by electrosensitive humans. Its also accounts for why using a mobile phone while driving makes you four times more likely to have an accident.

HOW CALCIUM CONTROLS METABOLISM

Calcium has a major role in maintaining membrane stability. The calcium concentration actually inside cells controls the rate of many metabolic processes. This includes the activity of many enzyme systems and the expression of genes. The concentration of calcium ions in the cytosol (the main part of the cell) is normally kept about a thousand times lower than that outside by metabolically-driven ion pumps in its membranes.

Many metabolic processes are then regulated by letting small amounts of calcium into the cytosol when needed. This is normally under very close metabolic control so that everything works at the right time and speed.

However, when electromagnetic exposure increases membrane leakiness, unregulated amounts of extra calcium can flood in. Just what happens then depends on how much gets in and what the cells are currently programmed to do. If they are growing, the rate of growth may be increased. If they are repairing themselves after injury, the rate of healing may be increased but if there is a mutant precancerous cell present, it may promote its growth into a tumour.

CALCIUM LEAKAGE AND BRAIN FUNCTION

Normal brain function in humans depends on the orderly transmission of signals through a mass of about 100 billion neurones. Neurons are typically highly branched nerve cells. They usually have one longbranch (the axon), which carries electrical signals as action potentials (nerve impulses) to or from other parts of the body or between relatively distant parts of the brain (a nerve contains many axons bundled together).

The shorter branches communicate with other neurons where their ends are adjacent at synapses. They transmit information across the synapses using a range of neurotransmitters. Neurotransmitters are chemicals secreted by one neuron and detected by the other. The exact patterns of transmission through this network of neurons is horrendously complex and determines our thoughts and virtually everything we do.

Calcium plays an essential role in this because a small amount of calcium must enter the neuron every time before it can release its neurotransmitters. Without it, the brain would be effectively dead. But what would happen if electromagnetically induced membrane leakage let in too much calcium?

One effect would be to increase the background level of calcium in the neurons so that they release their neurotransmitters sooner. This improves our reaction time to simple stimuli (which has been experimentally proven). However, it can also trigger the spontaneous release of neurotransmitters to transmit spurious signals have no right to be there. This feeds the brain false information.

Similar spurious action potentials may also be triggered in other parts of the neuron if leaks in the membrane temporarily short-circuit the normal voltage between its inside and outside. These unprogrammed action potentials will degrade the signal to noise ratio of the brain and reduce its ability to make accurate judgements.

It is technically difficult to detect these stray action potentials experimentally since they look like random noise in the measuring system. They and would in any case be swamped by the relatively strong electromagnetic signals used to induce them. However, similar spurious action potentials should be detectable if we removed some of structural calcium from the membrane by some other means.

One way to do this is to lower the concentration of calcium ions in the surrounding medium. For example, Matthews (1986) reported that exposing nerve and muscle cells to calcium concentration about 10-20 percent below normal made them significantly more excitable. This fits with our hypothesis.

These findings also explain many of the symptoms of hypocalcemia (alias hypocalcaemia). Hypocalcemia is a medical condition, usually caused by a hormone imbalance. In this condition the concentration of ionised calcium in the blood is abnormally low. By removing bound calcium from cell membranes, it should (and does) give similar effects to electromagnetism.

EXPERIMENTAL RESULTS

Experimental methods of detecting endogenous ac electrical oscillations in cells are developed in the works of C. Smith, S. Webb, H. Pohl, etc., by directly measuring the dielectrophoresis (the action of non-uniform electric ac fields on neutral particles). In order to study the signals detected by Smith and Pohl in yeast cells in more detail, a measuring system with an improved signal-to-noise ratio using a high impedance preamplifier for electronic detection of the endogenous ac fields was developed by R. Hölzel.

Discrete bands of endogenous oscillations in the range of 1.5 MHz to 34.8 MHz with the amplitudes of 0.5 - 7.0 mV in various yeast cells were detected. These studies revealed that the endogenous fields are strongest when cell metabolism is most active. No signs of ac oscillations are found in dead or heavily poisoned cells.

The measurements of ac electric fields around cells made it possible for H. Pohl to assume that endogenous oscillations must accompany cellular reproduction, and vice versa. The reproductive process cannot proceed without endogenous ac oscillations. This may be due to the deep involvement of electromagnetic field interactions (within and between cells and organisms) in metabolic energy exchange and transformation.

At the microscopic level, numerous attempts to elucidate the extremely-low-frequency (ELF) signal transduction pathways of the interactions with cell membranes and subcellular components were made. This was done by measuring various cellular and subcellular characteristics while exposing the studied systems to experimentally generated external ELF fields. Physically, magnetic field exposure results in an internal magnetic field, internal electric fields and internal currents.

FREQUENCY DEPENDANT VARIATIONS

It is noteworthy that observable responses of biological systems to electromagnetic field treatment occur from altering field exposure. No marked difference in various cell characteristics (rate of cell proliferation, histogram of the nuclear DNA content, rates of lactate production and glucose consumption and the ATP content) of exposed and intact cells is obtained by using static magnetic fields . Moreover, many processes turn out to be frequency dependent with thresholds or some peculiarities at certain values of external fields.

ELECTROMAGNETIC WINDOWS

Information obtained thus far is still insufficient to offer a reasonable mechanism for EMF interaction with biological tissue. Nevertheless, we would like to emphasise some general features of such kinds of interactions like frequency and power windowing in tissue interactions with weak EMF. W. Adey realised the windowing effect after studying the behavioural and neurophysiological effects of extremely low frequency (ELF) and modulated radio frequency (RF) fields as well as the responses of calcium ion binding in tissues to ELF and RF fields .

ALTERNATING CURRENTS AND BIOCHEMISTRY

Alternating currents are shown to affect also ion transport and ATP splitting via changes in the activation of the membrane of Na, K-ATPase. Both processes vary with frequency, and can be explained by the effects of the ionic currents on ion binding at the enzyme's active sites. These could account for the effects of EM fields on cells, as the transmembrane enzyme can convey the effect of an extracellular signal into the cell via ionic fluxes.

The works of M. Blank and co-authors from Columbia University (New York) deal with EM field effects on protein synthesis and the Na, K-ATPase function in cells. There has been debate on the biological effects from environmental, non-ionising electromagnetic waves. The probably the strongest experimental evidence has come from studies of changes in biosynthesis.

The magnetic field, or more likely induced currents, may effect certain processes when the DNA is already unravelled to some extent. Also in the process of forming messenger RNA (mRNA). An effect may also occur in the cell when this message is read and different amino acids are being added to the growing protein chain. Changes occur at the level of transcription (formation of mRNA) or translation where the message is made into a protein.

ELF FIELDS AND GENE EXPRESSION

Results of studies on the above show that the steady state levels of some RNA transcripts are increased when cells are exposed to ELF electric or magnetic fields. Experiments have exposed a variety of cell types, including dipteran salivary gland cells, yeast and human HL-60 cells. These data suggested that the physiological mechanism involved in the cellular response to ELF electromagnetic fields may be similar to or mimic the response to heat shock and that one effect of EM fields is directly at the transcription level.

As expected, the transcription response was similar to the generalised response of cells to stress (these results point to a possible link between EM field exposure and malignancy through the over-expression of stress genes and increases in stress proteins). Both electric and magnetic fields appear to stimulate the same genes.

The authors (S. G. Boxer, C. E. D. Chidsey, M. G. Roelofs) of the study also support the idea that integral membrane enzymes may couple to the electric field vector of an alternating electromagnetic field. This is limited to membrane-bound proteins that undergo large conformational changes during catalysis. However, the results in show that an intact membrane is not an absolute requirement for or transducing magnetic bio-effects.

CONTINUOUS WAVES CAN ALSO WORK

This is probably because living cells can introduce their own time variation in field strength. The membrane systems in active living cells are constantly on the move, e.g. from the Brownian motion of membrane-bound particles (a purely physical process due to molecular bombardment). Also in physiological processes such as their active transport. This exposes any given section of their membranes to a full frontal attack by the field in one orientation followed by a much quieter period if it rotates through 90 degrees and receives the signal edge-on.

This means that it experiences what looks (to it) like a time-varying field and may therefore give a physiological response even to a constant radio signal. However, because these are random changes and are not sharply pulsed, we might expect them to need stronger fields and/or longer exposure times if they are to give effects.

This may explain the unpleasant symptoms experienced by many electrosensitive individuals when using UMTS (3G) handsets or living close to high power TETRA base stations. Although neither signal is pulsed, the sheer proximity of the UMTS handset to the user and the raw power of a nearby TETRA base station may give the necessary signal strength. In addition, the lack of any quiet gaps in the signal increases the net exposure time, which may more than compensate for the lack of pulses.

HOW CALCIUM LOSS MAKES HOLES IN MEMBRANES

Cell membranes are made of sheets of fatty materials called phospholipids surrounding islands of protein. The proteins have a variety of metabolic functions, but the main role of the phospholipids is to fill the spaces between them and act as a barrier to prevent leakage. Calcium loss weakens the phospholipid sheet and makes it more likely to leak;

Fortunately, these pores are usually self-healing and the damage to the membrane is not permanent. However, during electromagnetic exposure there will be more tears, slower repair and consequently more overall leakage. The metabolic effects of even a brief period of leakage may be much longer lasting (e.g. if dormant genes are activated) and perhaps (as in the case of DNA damage) permanent.

DEFENSE MECHANISMS

Calcium pumps: Cells have to be able to pump out any extra calcium that has entered their cytosols. This is to reset the low cytosolic calcium level every time it is disturbed by a programmed calcium influx. They should therefore be able to respond to unprogrammed calcium influx due to electromagnetic exposure. This should minimise any unwanted metabolic effects, but the scope to do this is limited. If it were too effective, it would also prevent legitimate cell signalling.

Gap junction closure: If calcium extrusion fails and there is a large rise in internal calcium, it triggers the isolation of the cell concerned by the closure of its gap junctions (tiny strands of cytoplasm that normally connect adjacent cells) (Alberts et al. 2002). This also limits the flow of eddy currents through the tissue and so reduces the effects of radiation.

Heat shock proteins: These were first discovered after exposing cells to heat. They are also produced in response to a wide variety of other stresses, including weak electromagnetic fields. They are normally produced within minutes of the onset of the stress. They combine with the cell's enzymes to protect them from damage and shut down non-essential metabolism (the equivalent of running a computer in 'safe mode').

When the production of heat shock proteins is triggered electromagnetically it needs 100 million million times less energy than when triggered by heat, so the effect is truly non thermal (Blank & Goodman 2000). Their production in response to electromagnetic fields is activated by special base sequences (the nCTCTn motif) in the DNA of their genes.

When exposed to electromagnetic fields, they initiate the gene's transcription to form RNA, which is the first stage in the synthesis of the protein (Lin et al. 2001). As we can see, there are several defence mechanisms against damage by electromagnetic fields and there may be more we do not know about.

They probably evolved in response to natural electromagnetic fields such as those generated by thunderstorms. However they are now having their work cut out to respond to the continuous and all-pervading fields associated with modern living.

How well they perform will depend on many factors, including environmental conditions, the physiological condition of the cells and how much energy they have to spare. Consequently, they do not always succeed.

When the defences fail, we may get visible symptoms from the radiation, but when they succeed, there may be little obvious effect. The power and mobile phone companies have seized upon this characteristic variability to discredit work on the non-thermal effects of electromagnetic fields as being due to the experimental error. Nothing could be further from the truth.

ELECTROSENSITIVITY AND HYPOCALCEMIA - A POSSIBLE CURE

Symptoms of hypocalcemia include skin disorders, parethesia (pins and needles, numbness, sensations of burning etc.) fatigue, muscle cramps, cardiac arrhythmia, gastro-intestinal problems and many others. The symptoms of hypocalcemia are remarkably similar to those of electrosensitivity.

If you think you may be electrosensitive, how many of these do you have? If you have any of them, it may be worth having your blood checked for ionised calcium. It is possible that at least some forms of electrosensitivity could be due to the victims having their natural blood calcium levels bordering on hypocalcemia.

Electromagnetic exposure would then remove even more calcium from their cell membranes. This may push them over the edge and give them symptoms of hypocalcemia. If this is correct, conventional treatment for hypocalcaemia may relieve some if not all of these symptoms.

CONCLUSION

Many scientists believe the cell membrane plays a leading role in the interaction mechanisms between EMF and biological matter. The European Parliament Resolution B3-0280/92, clauses D and E, bases its concern on the matter of EMF health effects, in part, on recognition that the cell membrane is the primary site of cellular interaction of EMF and living tissues:

D. whereas, according to an increasing number of epidemiological and experimental studies, even slight exposure to non-ionizing electromagnetic fields increases the risk of cancer, can be accompanied by nervous disorders and disruption of the circadian rhythm and seems capable of affecting developing organisms,

E. whereas the results of many in vivo and in vitro studies show increasingly clearly the interaction mechanisms underlying such disorders and illnesses, centered mainly in the cell membrane, lead to disruption of melatonin secretions, ornithine decarboxylase activity and T lymphocyte efficacy, testifying to the probable role of non-ionizing radiation in promoting cancer.

The first step of course is to become aware of how much they are impacting you. Scientists working in the field of "bioelectromagnetics" are now convinced that manmade EMFs disturb biological processes (induce biological effects).

Some of the biological effects seen in the laboratories are similar to biochemical mechanisms believed to be responsible for neurological effects like short term memory loss, whereas others are believed to be involved in the development of serious disorders like cancer, Alzheimer's and Parkinson's.

REFERENCES

Adey WR: Frequency and Power Windowing in Tissue Interactions with Weak Electromagnetic Fields. Proceedings of the IEEE, 1980, vol. 63, no. 1, p. 119-125.

Adey WR: "Tissue interactions with non-ionizing electromagnetic fields", Physiol. Rev., 1981, vol. 61, p: 435.

Adey WR (1990) 'Electromagnetic fields, cell membrane amplification, and cancer promotion'. In: Wilson BW, Stevens RG, Anderson LE (eds) Extremely Low Frequency Electromagnetic Fields: the Question of Cancer . Battelle Press, Columbus, Ohio, pp 211-249

A. F. Lawrence, W. R. Adey: Nonlinear Wave Mechanisms in Interactions between Excitable Tissue and Electromagnetic Fields, Neurological Research, v. 4, n. 1/2, 1982, p. 115-153.

Bawin SM, Adey WR (1976) 'Sensitivity of calcium binding in cerebral tissue to weak environmental electric fields oscillating at low frequency'. Proc Nat Acad Sci USA 73: 1999-2003

Blackman CF (1990) 'ELF effects on calcium homeostasis'. In: Wilson BW, Stevens RG, Anderson LE (eds) Extremely Low Frequency Electromagnetic Fields: the Question of Cancer . Battelle Press, Columbus, Ohio, pp 189-208

Blackman CF, Benane SG, Kinney LS, House DE, Joines WT (1982) 'Effects of ELF fields on calcium-ion efflux from brain tissue in vitro'. Radiat. Res. 92: 510-520

Blank M., Soo L. (1992): Threshold for Inhibition of Na, K-ATPase by ELF Alternating Currents. Bioelectromagnetics, v. 13, p. 329-333.

Blank M., Soo L. (1993): The Na, K-ATPase as a model for electromagnetic field effects on cells. Bioelectrochem. and Bioenergetics, v. 30, p. 85-92.

Blank M., L. Soo, H. Lin, A. S. Henderson, R. Goodman (1992): Changes in transcription in HL-60 cells following exposure to alternating currents from electric fields. Bioelectrochemistry and Bioenergetics, v. 28, p. 301-309.

Blank M., O. Khorkova, R. Goodman (1994): Changes in polypeptide distribution stimulated by different levels of electromagnetic and thermal stress, Bioelectrochemistry and Bioenergetics, v. 33, p. 109-114.

Blank M, Goodman R (2000) 'Stimulation of stress response by low frequency electromagnetic fields: possibility of direct interaction with DNA'. IEEE Trans Plasma Sci 28: 168-172

Blank M, Goodman R, Electromagnetic fields stress living cells, Pathophysiology 16 (2009) 71–78,

Blank M. (1992): Na, K-ATPase function in alternating electric fields, FASEB Journal, v. 6, p.2434-2438.

Blank M (1993): Biological effects of electromagnetic fields, Bioelectrochemistry and Bioenergetics, v. 32, p. 203-210.

Diem E, Schwarz C, Aldkofer F, Jahn O, Rudiger H (2005) 'Non-thermal DNA breakage by mobile phone radiation (1800 MHz) in human fibroblasts and in transformed GFSH-R17 rat granulosa cells in vitro'. Mutation Research/Genetic Toxicology and Environmental Mutagenesis 583: 178-183

Garssen J, Goettsch W, Slob FR, De Gruijl FR, Van Loveren H. Risk assessment of UVB effects on the resistance to infectious diseases. Photochem Photobiol 1995; 61S: 46S.

Goldsworthy A (2006) 'Effects of electrical and electromagnetic fields on plants and related topics'. In: Volkov AG (ed.) Plant Electrophysiology - Theory & Methods . Springer-Verlag Berlin Heidelberg 2006. Pp 247-267.

Goodman R., Henderson A. S. (1988): Exposure of salivary gland cells to low frequency electromagnetic fields alters polypeptide synthesis, Proc.Natl.Acad.Sci., USA, v. 85, p. 3928-3932.

Ha B-Y (2001) 'Stabilization and destabilization of cell membranes by multivalent ions'. Phys. Rev . E. 64: 051902 (5 pages)

H. A. Pohl: Dielectrophoresis: The Behavior of Matter in Non-uniform Electric Fields, Cambridge University Press, 1978.

H. A. Pohl: Natural oscillating fields of cells, In: Coherent excitations in Biological Systems. (H.Frölich, F. Kremer, eds.), Springer Verlag, Berlin, 1983, 199-210.

H. A. Pohl: Natural ac electric fields in and about cells, Phenomena, v. 5, 1983, p. 87-103.

H. A. Pohl: Natural electrical RF oscillation from cells, J. Bioenerg. Biomembr., 1981, vol. 13, p.149-169.

Liboff AR, McLeod BR, Smith SD (1990) 'Ion cyclotron resonance effects of ELF fields in biological systems'. In: Wilson BW, Stevens RG, Anderson LE (eds) Extremely Low Frequency Electromagnetic Fields: the Question of Cancer . Battelle Press, Columbus, Ohio, pp 251-289

Lin H, Blank M, Rossol-Haseroth K, Goodman R (2001) 'Regulating genes with electromagnetic response elements'. J Cellular Biochem 81: 143-148

Lodish H, Berk A, Matsudaira P, Kaiser CA, Krieger M, Scott MP, Zipursky SL, Darnell J. *Molecular Cell Biology, 5th, New York: WH Freeman and Company, 197–234.*

Matthews EK (1986) 'Calcium and membrane permeability'. British Medical Bulletin 42: 391-397

McLeod BR, Smith SD, Liboff AR (1987) 'Potassium and calcium cyclotron resonance curves and harmonics in diatoms (A. coffeaeformis)'. J Bioelectr 6: 153-168

Mehedintu M, Berg H (1997) 'Proliferation response of yeast Saccharomyces cerevisiae on electromagnetic field parameters'. Bioelectrochem Bioenerg 43: 67-70

Melikov KC, Frolov VA, Shcherbakov A, Samsonov AV, Chizmadzhev YA, Chernomordik LV (2001) 'Voltage-induced nonconductive prepores and metastable single pores in unmodified planar lipid bilayer'. Biophys J 80: 1829-1836

Mevissen M., Stamm A., Buntenkötter S., Zwingelberg R., Wahnschaffe U., Löscher W. (1993): Effects of Magnetic Fields on Mammary Tumor Development Induced by 7,12-Dimethylbenz(a)anthracene in Rats. Bioelectromagnetics, v. 14, p. 131-143.

R Ader and N Cohen. Behaviorally conditioned immunosuppression. Psychosomatic Medicine, Vol 37, Issue 4 333-340

R. Choy, J. A. Monro, C. W. Smith: Electrical sensitivities in Allergy Patients, Clinical Ecology, 1987, vol. 4, n. 3, p. 93-102.

R. Goodman, M. Blank, H. Lin, R. Dai, O. Khorkova, L. Soo, D. Weisbrot, A. Henderson (1994): Increased levels of hsp70 transcripts induced when cells are exposed to low frequency electromagnetic fields. Bioelectrochemistry and Bioenergetics, v. 33, p. 115-120.

S. G. Boxer, C. E. D. Chidsey, M. G. Roelofs (1982): Use of large magnetic field to probe photo-induced electron transfer reaction: an example from photosynthetic reaction center. J. Amer. Chem. Soc., v.

104, No. 5, p. 1452-1454.

Schreier N, Huss A, Roosli M. The prevalence of symptoms attributed to electromagnetic field exposure: a cross-sectional representative survey in Switzerland. Soz Praventivmed. 2006;51(4):202-9.

Serpersu E. H., Tsong T. Y. (1984): Activation of electrogenic Rb+ transport of Na,K-ATPase by an electric field. J.Biol.Chem., v. 259, p. 7155-7162.

Smith SD. McLeod BR, Liboff AR (1993) 'Effects of SR tuning 60Hz magnetic fields on sprouting and early growth of Raphanus sativus'. Bioelectrochem Bioenerg 32: 67-76

Stenz H-G, Wohlwend B, Weisenseel MH (1998) 'Weak AC electric fields promote root growth and ER abundance of root cap cells'. Bioelectrochem Bioenerg 44: 261-269

Yamaguchi H., Hosokawa K., Soda A., Mizamoto H., Kinouchi Y. (1993): Effects of seven months exposure to a static 0.2 T magnetic field on growth and glycolytic activity of human gingival fibroblasts, Biochem. Biophys. Acta, v. 1156, p. 302-306.

Yariktas M, Doner F, Ozguner F, Gokalp O, Dogrutt H, Delibas N. Nitric oxide level in the nasal and sinus mucosa after exposure to electromagnetic field. Otolaryngol Head Neck Surg. 2005 May;132(5):713-6.

6. WHAT SECRETS DOES THE US ARMY KNOW ABOUT RADIATION?

After many years of seeing patients with radiation sensitivity and damage, I became sensitized to this. It has come to a point where whenever I am seeing a patient who seems to have strange or bizarre symptoms, I tend to suspect radiation damage. So I developed a series of protocols to identify this reliably.

As we do a literature review of the harmful effects of EMR, you will notice that often multiple systems of the body seem to be involved. When EMR damage occurs concurrently with heavy metal toxicity (mercury, lead or cadmium for example) or exposure to organic toxins (benzene, xylene or toluene), the medical condition is often hard to diagnose, manage or treat.

OVERVIEW OF EMF NON-THERMAL EFFECTS

What you need to know

Scientists working in the field of "bioelectromagnetics" are now convinced that man-made EMFs disturb biological processes (induce biological effects). These disturbances can trigger a variety of conditions like miscarriage, birth defects, breast cancer (in both men and women), adult and childhood leukemia, hormone suppression, Alzheimers disease, Parkinsons disease, Lou Gehrigs disease and ALS.

Since World War II, the U.S. Army has been the world's largest user of electrical, electronic, and telecommunication equipment and was the first to recognize effects induced by EMFs on military personnel (radar operators on WW II battleships).

The background for the project was a concern of the U.S. Army about the biological effects and potential health consequences to personnel exposed to EMFs.

The technology to protect against radiation was developed in a large, on-going research project at the Catholic University of America (CUA), Department of Physics in Washington, D.C.

> The research showed that that the right technology can be designed to prevent non thermal biological effects of EMF exposures. When the biological effects are prevented, there cannot possibly be any health effects associated with non-thermal exposures.
>
> The technology works by blocking the EMF-induced physical-chemical interactions around the cell membrane receptors, probably by preventing cooperative sensing from happening. In this way, the cascade of biochemical events which would result inside the cell as a function of membrane receptors sensing the constant, oscillating EMF, is prevented from being triggered. The technology re-establishes the situation around the cell as if it was non-exposed.
>
> To date, a total of eight studies have pointed to the protective effects of melatonin and vitamins C and E in stemming the damage caused by cell phone emissions.

In response to a growing body of scientific evidence, the existence of health and biological effects associated with exposures to EMFs is becoming more widely known and accepted. More and more scientists now believe that the existence of significant non-thermal effects induced by low-level non-ionizing EMF are a reality.

For example, on Wednesday, July 24, 1998, a 28 member panel convened by the National Institute of Environmental Health Sciences (NIEHS) decided that extremely low frequency (ELF) electromagnetic fields should be regarded as possible carcinogens.

The final vote of the panel was 19 to 9 in favor of categorizing ELF EMFs, such as those from power lines and electrical appliances, as possible carcinogens. The vote followed one year of study including three major, multi-day symposia and a final 10 day intensive meeting of scientists to review and debate the scientific and medical literature.

In October 1998 at the University of Vienna Workshop on Possible Biological and Health Effects of Radio Frequency (RF/MW) Electromagnetic Fields, the following resolution was made by the participating scientists (the "Vienna Resolution"):

"The participants agreed that biological effects from low-intensity exposures are scientifically established. However, the current state of scientific consensus is inadequate to derive reliable exposure standards. The existing evidence demands an increase in the research efforts on possible health impact and on adequate exposure and dose assessment."

In conclusion, scientists working in the field of "bioelectromagnetics" are now convinced that man-made EMFs disturb biological processes (induce biological effects). Some of the biological effects seen in the laboratories are similar to biochemical mechanisms believed to be responsible for neurological effects like short term memory loss, whereas others are believed to be involved in the development of serious disorders like cancer, Alzheimer's and Parkinson's.

However, the conclusion mentioned above, formulated by the 28 person scientific work group under NIEHS stating that EMFs should be regarded as possible carcinogens, is a warning; already, a substantial body of scientific evidence do point to a relation between EMFs and cancer. Below is shown a summary of scientific findings on EMF-induced biological effects related to the cascade of events believed to be significant in the development of cancer:

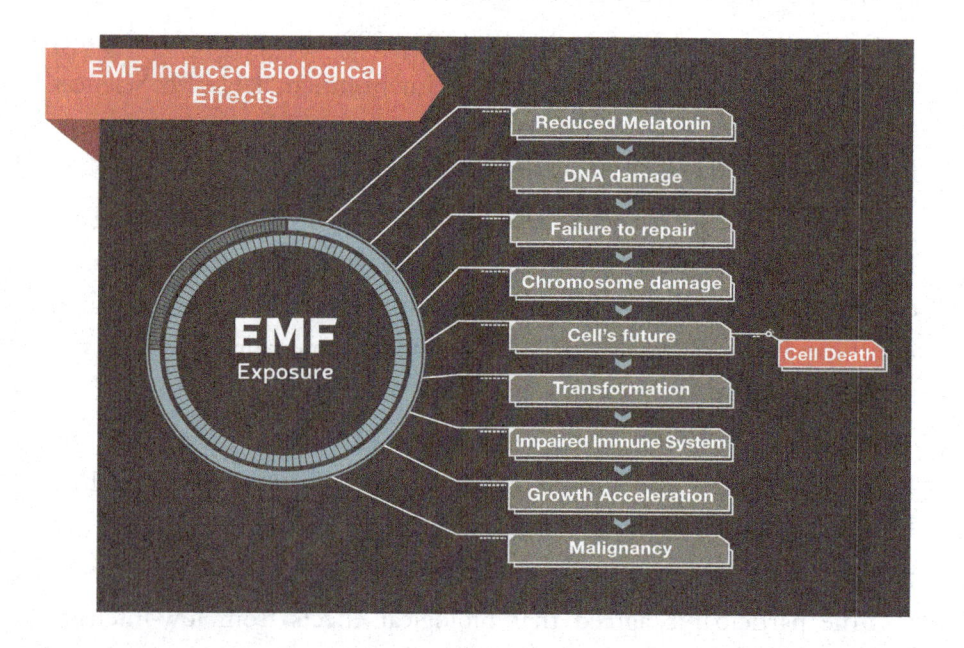

Fig.9: EMF induced biological effects

Below is shown a summary of the biological effects which have been found in peer-reviewed and published scientific studies. It is also shown how these biological effects may lead to physiological changes. This summary is not exhaustive, and much more research will be needed in the future to expand and refine the picture.

Summary of Scientific Findings:

I. Physico-Chemical Events

ELF FIELDS → FIELD AMPLIFICATION. COOPERATIVE SENSING.

RF / MW FIELDS → STRONG INDUCED FIELD. COOPERATIVE SENSING.

↓

CHANGED ELECTROMAGNETIC MICROENVIRONMENT AND CHARGE DISTRIBUTION AND CELL. MEMBRANE RECEPTORS AFFECTED COOPERATIVELY CANCEL OUT THERMAL NOISE EFFECTS

↓

EFFECT ON TRANSMEMBRANE ION PUMPING MECHANISMS LEAD TO CHANGED ION DISTRIBUTION INSIDE AND OUTSIDE CELL.

↓

CHANGED ION CONCENTRATION INSIDE CELLS, NOTABLY CALCIUM, AFFECT ENZYME-DRIVEN INTRA-AND INTERCELLULAR MESSENGER SYSTEMS

↓

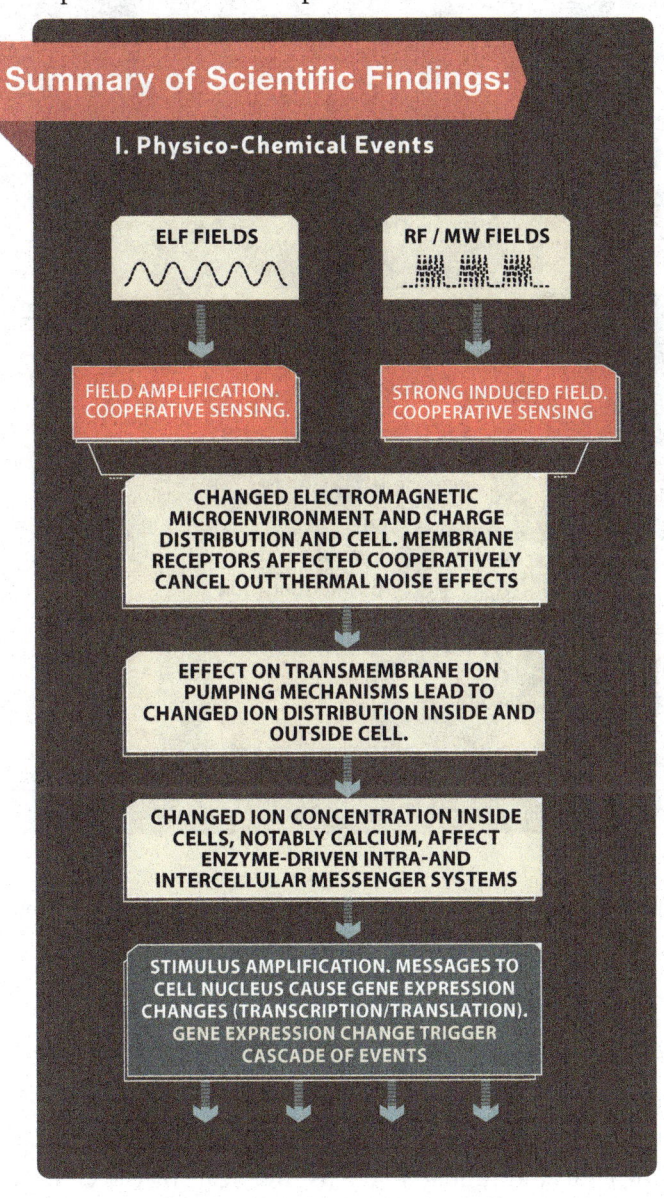

STIMULUS AMPLIFICATION. MESSAGES TO CELL NUCLEUS CAUSE GENE EXPRESSION CHANGES (TRANSCRIPTION/TRANSLATION). GENE EXPRESSION CHANGE TRIGGER CASCADE OF EVENTS

Fig. 10: Summary of scientific findings of ELF, RF/MW fields on cells

Summary of Scientific Findings:
II. The Biochemical › Biological › Physiological Effects

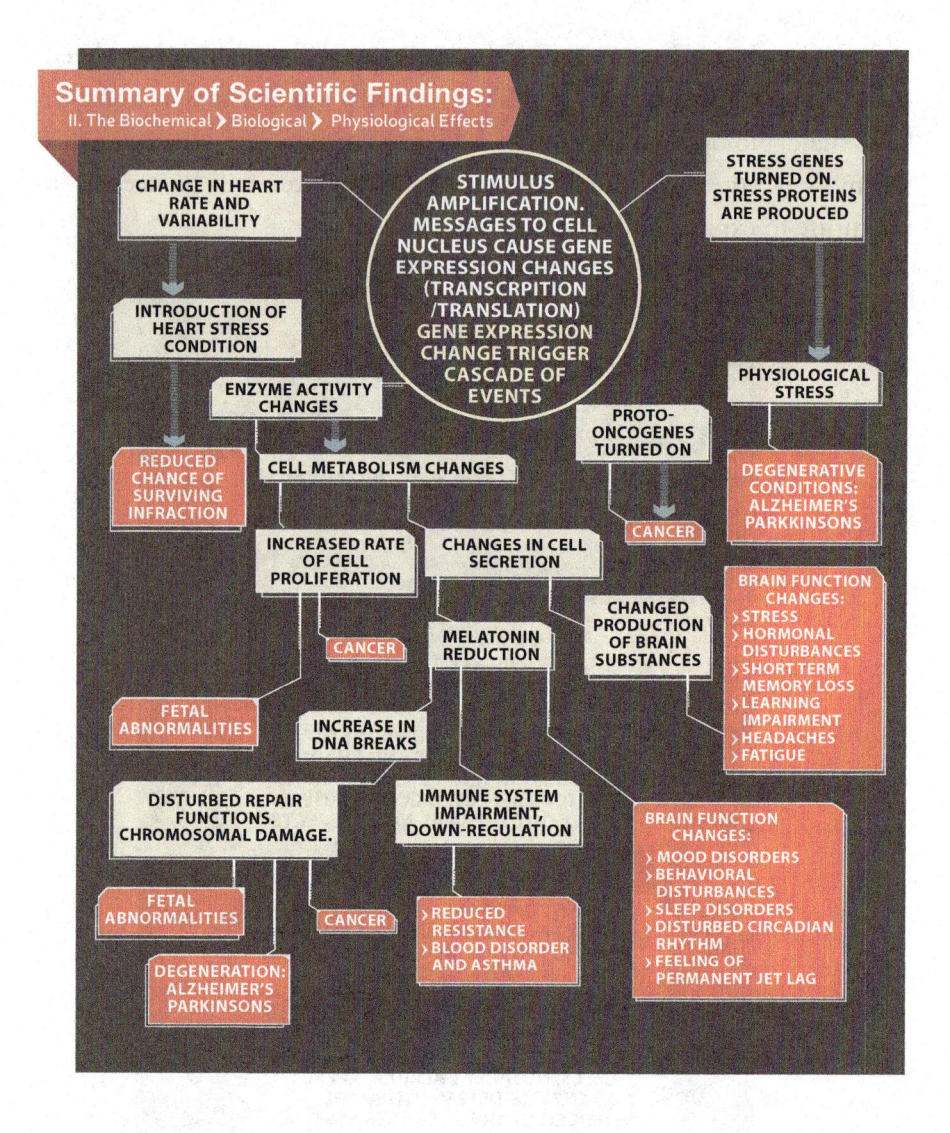

- STIMULUS AMPLIFICATION. MESSAGES TO CELL NUCLEUS CAUSE GENE EXPRESSION CHANGES (TRANSCRPITION/TRANSLATION) GENE EXPRESSION CHANGE TRIGGER CASCADE OF EVENTS
- CHANGE IN HEART RATE AND VARIABILITY → INTRODUCTION OF HEART STRESS CONDITION
- ENZYME ACTIVITY CHANGES → REDUCED CHANCE OF SURVIVING INFRACTION
- CELL METABOLISM CHANGES
  - INCREASED RATE OF CELL PROLIFERATION → CANCER; FETAL ABNORMALITIES
  - CHANGES IN CELL SECRETION
    - MELATONIN REDUCTION → INCREASE IN DNA BREAKS → DISTURBED REPAIR FUNCTIONS. CHROMOSOMAL DAMAGE. → FETAL ABNORMALITIES; CANCER; DEGENERATION: ALZHEIMER'S PARKINSONS
    - CHANGED PRODUCTION OF BRAIN SUBSTANCES → BRAIN FUNCTION CHANGES: ›STRESS ›HORMONAL DISTURBANCES ›SHORT TERM MEMORY LOSS ›LEARNING IMPAIRMENT ›HEADACHES ›FATIGUE
  - IMMUNE SYSTEM IMPAIRMENT, DOWN-REGULATION → ›REDUCED RESISTANCE ›BLOOD DISORDER AND ASTHMA
- PROTO-ONCOGENES TURNED ON → CANCER
- STRESS GENES TURNED ON. STRESS PROTEINS ARE PRODUCED → PHYSIOLOGICAL STRESS → DEGENERATIVE CONDITIONS: ALZHEIMER'S PARKINSONS; BRAIN FUNCTION CHANGES: ›MOOD DISORDERS ›BEHAVIORAL DISTURBANCES ›SLEEP DISORDERS ›DISTURBED CIRCADIAN RHYTHM ›FEELING OF PERMANENT JET LAG

Fig. 11: Biochemical/biological/physiological events conditions and potential disease

ARMY DEVELOPS NOISE FIELD TECHNOLOGY

The technology to protect against radiation was developed in a large, on-going research project at the Catholic University of America (CUA), Department of Physics in Washington, D.C. The research project was initiated in 1986 and funded in its first five years by the U.S. Army Walter Reed Army Institute Department of Microwave Engineering.

The background for the project was a concern of the U.S. Army about the biological effects and potential health consequences to personnel exposed to EMFs. Since World War II, the U.S. Army has been the world's largest user of electrical, electronic, and telecommunication equipment and was the first to recognize effects induced by EMFs on military personnel (radar operators on WW II battleships).

The research showed that that the right technology can be designed to prevent non thermal biological effects of EMF exposures. When the biological effects are prevented, there cannot possibly be any health effects associated with non-thermal exposures. The technology works by blocking the EMF-induced physical-chemical interactions around the cell membrane receptors, probably by preventing cooperative sensing from happening. In this way, the cascade of biochemical events which would result inside the cell as a function of membrane receptors sensing the constant, oscillating EMF, is prevented from being triggered. The technology re-establishes the situation around the cell as if it was non-exposed. This is illustrated in Figure 17.

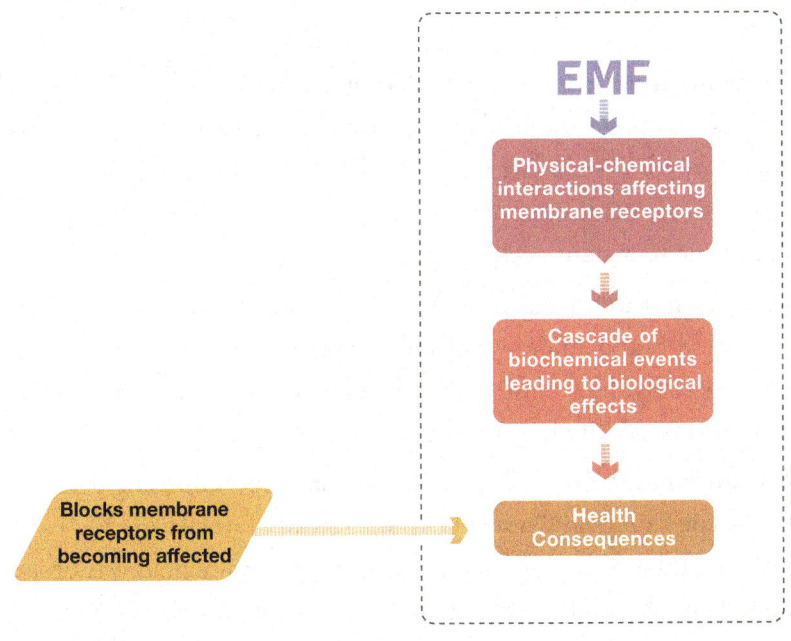

Fig. 12: How the EMF technology blocks harmful biological effects of EMR

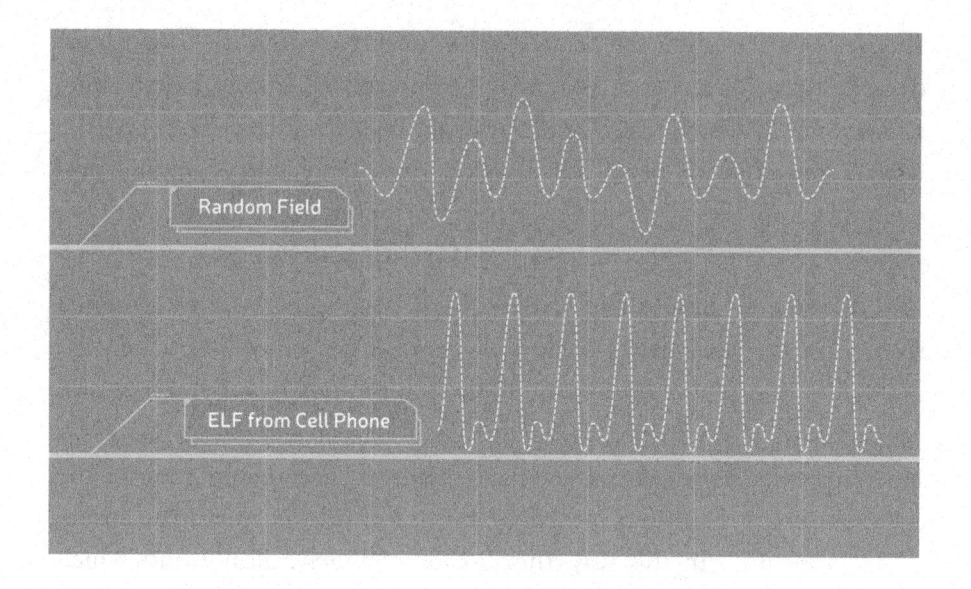

Fig. 13: The time variance of a natural EMF frequency, amplitude and waveform varies at random (top). A "man-made" EMF, emitted from the internal circuitry of a digital mobile telephone. Frequency, amplitude and waveform are constant for a certain time period (bottom).

The technology works by simulating nature's random EMFs. The research team at CUA found that the requirement for biological events to be induced by EMF exposure is that the EMF – apart from being spatially coherent like all fields external to the body – is steady for a minimum period of time (constant oscillation) .

The CUA research team found – like other scientists working in this area – that ELF and microwaves induced remarkably similar biological effects. Especially ELF and ELF modulated microwaves are virtually identical in their ability to induce biological effects.

Most important: The scientists found that superimposing a random, "natural"-type ELF field ("noise" field) on a steadily oscillating EMF could mitigate the induced biological effects; and if the noise field was properly engineered, any biological effect tested was completely eliminated.

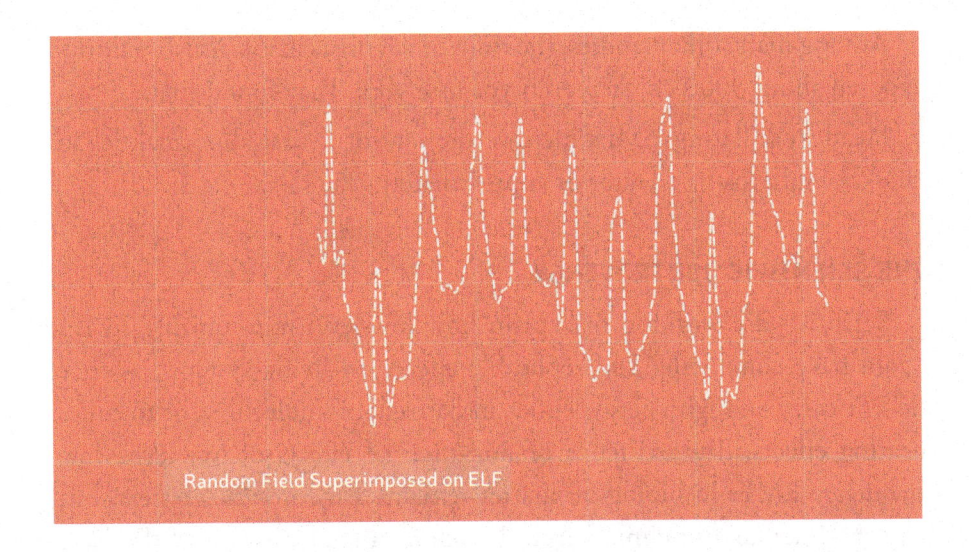

Fig. 14: Noise field superimposed on the constant ELF emitted from the circuitry of digital mobile phone

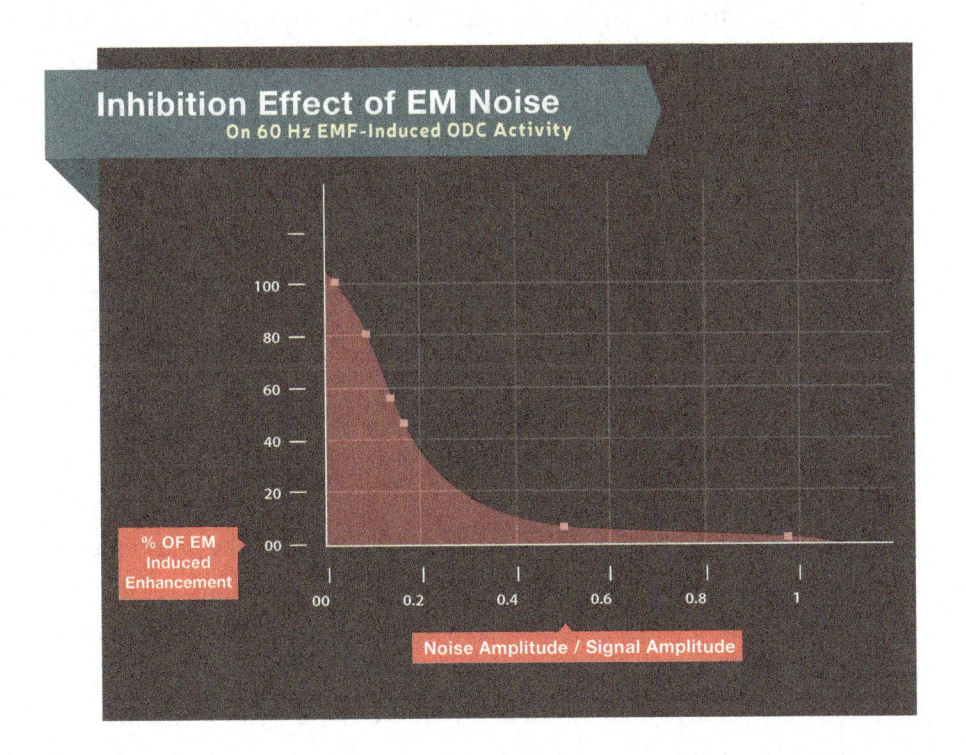

Fig. 15: EM noise inhibits the 60-Hz ELF-induced ODC activity in cells

Among many other studies, the team of scientists in CUA's laboratories exposed about 3000 chicken embryos to EMFs. They found that:

Steady, oscillating EMFs increase the rate of neural abnormalities in chicken embryos by a factor of approximately 2.5.

THE SCIENTIFIC CONTROVERSY

Scientists are working to explain how low-level non-ionizing EMFs could have convincing and replicable effects on living systems, as they do not carry enough energy, either to damage biomolecules, or to cause heating effects. The existence of non-thermal effects of low-level, low frequency EMFs in biological matter remains a theoretical mystery, but an experimental and clinical fact. Therefore, an explanation of the facts happening in the laboratories of experimental scientists, requires a new paradigm for the scientific perception of life processes.

The debate has so far been dominated by physical concepts and arguments. The theoretical "signal-to-noise" dilemma is among the most persistent of the criticisms which have cast doubt on the credibility of reports that weak ELF fields have been shown to affect life processes in biological systems.

Exogenous ELF fields reported to induce biological effects have been shown to be orders of magnitude weaker than the endogenous fields associated with "thermal noise" – the randomly fluctuating local fields on the surface of the cell membranes, caused by random thermic movements of charges (ions) in the inter- and intracellular liquids.

In other words, according to this theoretical argument, the induced electrical field of an exogenous ELF should be "drowned" by the thermal noise field inside the tissue. This belief is drawn from the solutions to fundamental mathematical equations from electromagnetic theory, according to which tissue is modeled as an ensemble of simple, smooth, non-interacting insulators surrounded by a conducting medium. These simplified equations result in an induced electrical field which has an order of magnitude equal to the exogenous field.

The simplified theory neglects the electromagnetic detail of the biological cell, which is a complex structure, rich with convoluted, charged surfaces. At the Catholic University of America (CUA) in Washington, D.C., Dr. Joan Farrell has made calculations taking into account a more realistic representation of the electromagnetic properties of the biological cell.

The result of this more accurate mathematical model lead to the discovery that exogenous ELF fields induce an electric field in tissue which may become amplified by orders of magnitude relative to the exogenous field. It allows for calculation of the fields close to the cell's surface, where the cell's array of "detectors" or chemoreceptors operate.

Compared to ELF fields, it is easier to accept that RF and MW fields may cause biological effects, since these fields carry more energy, and electric fields induced in tissue by RF/MW are – without enhancement mechanisms in play – in themselves orders of magnitude higher than the thermal noise fields. Still, however, many scientists are skeptical towards the existence of non-thermal biological effects of low-level RF/MW fields, as they believe that the only effects possible are associated with heating. These scientists are still stuck in the biochemical model of cellular damage.

Another factor adding to the skepticism towards the existence of biological effects of low-level non-ionizing EMF is the issue of replication. Some laboratories may see one level of effect of an EMF on a specific biological mechanism in a particular line of cells, tissue or animals, while other laboratories may find another level or even no effect.

Scientists at the Catholic University of America have studied this phenomenon and found that it may be due largely to differences in the genetics between different strains of cells or animals. Most laboratories usually do not control this confounder when attempting to replicate the results of another laboratory. So the scientists at the Catholic University of America found that the lack of reproducibility was due to poor experimental design.

NUTRITIONAL PROTECTION AGAINST CELL PHONE RADIATION

As growing evidence points to the potential adverse health impact of exposure to cell phone radiation, scientists are seeking strategies to prevent or mitigate these effects. Currently, nutritional researchers are exploring whether melatonin, vitamin C, and and vitamin E can ameliorate the detrimental effects caused by radiation emitted by cell phones.

To date, a total of eight studies have pointed to the protective effects of melatonin and vitamins C and E in stemming the damage caused by cell phone emissions. In particular, these agents show promise in averting the increased oxidative stress that is thought to contribute to an increased risk of certain cancers. These studies have unveiled statistically significant protective effects of melatonin and vitamins C and E against the effects of the radiation frequency at which cell phones emit and receive radio frequency radiation.

Six of these eight studies were controlled, short-term studies (ranging from 10-30 days) in rodents. Each study examined 24-30 subjects. Study subjects were divided equally into three groups: one group received radiation exposure; another received active treatment with melatonin only, vitamin C only, or vitamins C and E before radiation exposure; and a control group did not receive radiation or active treatment.

After the treatment period, scientists examined skin sections for radiation injury and analyzed blood and urine for markers of oxidative stress. They found significant kidney damage, skin changes, oxidative stress, and fibrosis in the animals who received radiation exposure only. Remarkably, these effects were reversed in the groups that received melatonin and vitamins C and E.

Another two controlled studies in rodents, one of 10 days' and another of 60 days' duration, revealed that melatonin significantly protects against retinal (eye) and kidney tissue damage caused by cell phone radiation, as compared with subjects that did not receive melatonin.

In one study, melatonin levels in the blood were measured in 226 male electric utility workers who were categorized according to cell phone use. The study concluded that workers who used cell phones for more than

25 minutes per day had decreased melatonin production and revealed a relationship between increased cell phone use and decreasing melatonin levels in the blood.

Yet six other studies, two in humans and four in rodents found that melatonin levels remained unchanged after radiation exposure. One human study did suggest that cell phone radiation may impact melatonin onset time. These were small studies, however, the majority of which were less than 28 days' duration.

Melatonin is a vital natural neurohormone (hormone secreted by or acting on a part of the nervous system) that acts as a potent free radical scavenger and antioxidant. Melatonin regulates the daily circadian rhythm and is essential to self-repair and regeneration. Given melatonin's protective effects, these findings warrant further research into the effect of cell phone radiation on melatonin in larger, longer-term, well-controlled human studies.

SCIENTIFIC ANALYSIS OF PROBLEM

Industry associations are eagerly trying to downplay the risks and frequently state that there is no proof of any health risks of EMF exposures. However, the fact is that:

Man made EMFs evidently induce significant biological effects in human cells, tissues and whole animals. These biological effects are linked to biochemical and physiological events believed to be associated with the development of adverse health conditions; whenever there is evidence of such interactions, there is justified reason for concern;

A majority of epidemiological studies point to an association between adverse health conditions and EMF exposures; this is another reason justifying public concern.

Added together, these two individual dimensions of "circumstantial evidence" brings a third dimension into the matter of concern. Therefore, the only statement which can be made for certain is:

"There is no scientific evidence that man made EMFs are safe".

The options available to consumers who are concerned about being "guinea pigs" in the large scale experiment set up by the electrical, electronics and telecommunications industries are the following:

- Prudent avoidance of the near field of systems and appliances emitting strong EMFs such as cellular phones and base stations, computers, hair dryers, power line systems, transformer stations, TV transmission towers, photocopying machines, food processors, microwave ovens, sewing machines, electric motors, etc.

This is often not practical, and consumers may not be knowledgeable about the proximity of transformer stations, underground power lines, cell phone towers, etc. In many occupation it is not possible to cut down the use of devices like cellular phones and computers. Besides, as research has shown, nobody knows what safe levels of EMF exposures are.

- Use of devices like the ear piece for mobile phones; however, many cell phone users place their handsets in their belts when they use their ear piece, thereby exposing large deposits of blood cells in the hip bone marrow or their genitals to the radiation. Furthermore, even at a distance, the radiation from a mobile phone antenna may induce significant biological effects.

There are physical shielding devices available in the market, especially for cell phones, and some of these sell well. The problems associated with shielding are the following:

- It is not practical to shield against ELF fields which penetrate virtually all materials except a special metal alloy (mu metal) which is as expensive as 24 ct gold; thus, a shield for a cell phone would do nothing to the ELF emission from the handset body;

- The shield may take away some of the microwave radiation of the cellular phone antenna; however, the level of the radiation will still be far above the thresholds for induction of biological effects which may to be thousands of times below the level actually emitted from the antenna;
- The shield will in many situations hamper the operation of the phone, depending on the direction to the nearest base station; if the head of the user is in the direction of the nearest base station, the phone will have to increase its output power to reach an alternative base station.
- Shielding devices may appear clumsy and unattractive.
- There are products being offered in the market consisting of little containers filled with "magnetized water", claimed to emit a protective magnetic field which eliminates potential hazards of the radiation. There is no published and peer-reviewed scientific backing for these claims.

SUMMARY

Dr. Neil Cherry of Lincoln University, New Zealand, in a recent report concluded:

"Scientific studies at the cellular level, whole animal level and involving human populations, show compelling and comprehensive evidence that RF/MW exposure down to very low levels, levels which are a minute fraction of present "safety standards", result in altered brain function, sleep disruption, depression, chronic fatigue, headache, impaired memory and learning, adverse reproductive outcomes including miscarriage, still birth, cot death, prematurely and birth deformities.

The US Army became concerned about the issue, did scientific studies and actually worked out a solution for the army personnel. If the US army was concerned, we should be as well. However the Noise Field Technology originally developed for the army was a power driven device. Unfortunately this technology was not made available to civilians. The next chapter is about amazing new technology that will protect us and our family against radiation.

REFERENCES

Catholic University of America: Simultaneous Application of a Spatially Coherent Noise Field Blocks Response of Cell Cultures to a 60Hz Electromagnetic Field (pdf)

Vitreous State Laboratory & Catholic University of America: Bioeffects Induced by Exposure to Microwaves are Mitigated by Superposition of ELF Noise (pdf)

Vitreous State Laboratory & Catholic University of America & Walter Reed Army Institue of Research: The Role of Coherence Time in the Effect of Microwaves on Ornithine Decarboxylase Activity (pdf)

University of Western Ontario: Effectiveness of Noise in Blocking Electromagnetic Effects on Enzyme Activity in the Chick Embryo (pdf)

Columbia University: Electric and Magnetic Noise Blocks the 60 Hz Magnetic Field Enhancement of Steady State c-myc Transcript Levels in Human Leukemia Cells (pdf)

Bioelectromagnetic Research Laboratory & University of Washington: Interaction of Microwaves and a Temporally Incoherent Magnetic Field on Spatial Learning in the Rat (pdf)

Ayata A, Mollaoglu H, Yilmaz HR, Akturk O, Ozguner F, Altuntas I. Oxidative stress-mediated skin damage in an experimental mobile phone model can be prevented by melatonin. J Dermatol. 2004 Nov;31(11):878-83.

Balci M, Devrim E, Durak I. Effects of mobile phones on oxidant/antioxidant balance in cornea and lens of rats. Curr Eye Res. 2007 Jan;32(1):21-5.

Brustolim D, Ribeiro-dos-Santos R, Kast RE, Altschuler EL, Soares MB. "A new chapter opens in anti-inflammatory treatments: the antidepressant bupropion lowers production of tumor necrosis factor-alpha and interferon-gamma in mice." Int Immunopharmacol. 2006 Jun;6(6):903-7

Burch, J.B et al "Cellular telephone use and excretion of a urinary melatonin metabolite". In: Annual review of Research in Biological Effects of electric and magnetic fields from the generation, delivery and use of electricity, San Diego, CA, Nov. 9-13, P-52.

Burch JB, Reif JS, Noonan CW et al. Melatonin metabolite excretion among cellular telephone users. Int J Radiat Biol. 2002 Nov;78(11):1029-36.

De Seze R, Ayoub J, Peray P, Miro L, Touitou Y. Evaluation in humans of the effects of radiocellular telephones on the circadian patterns of melatonin secretion, a chronobiological rhythm marker. J Pineal Res 1999 Nov;27(4):237-42.

Di Lullo GA, Sweeney SM, Korkko J, Ala-Kokko L, San Antonio JD (2002). "Mapping the ligand-binding sites and disease-associated mutations on the most abundant protein in the human, type I collagen". J. Biol. Chem. 277 (6): 4223–31.

Hata K, Yamaguchi H, Tsurita G, et al. Short term exposure to 1439 MHz pulsed TDMA field does not alter melatonin synthesis in rats. Bioelectromagnetics. 2005 Jan;26(1):49-53.

Jones CK, Eastwood BJ, Need AB, Shannon HE. Analgesic effects of serotonergic, noradrenergic or dual reuptake inhibitors in the carrageenan test in rats: evidence for synergism between serotonergic and noradrenergic reuptake inhibition. Neuropharmacology. 2006 Dec;51(7-8):1172-80.

Koyu A, Ozguner F, Cesur F, et al. No effects of 900 MHz and 1800 MHz electromagnetic field emitted from cellular phone on nocturnal serum melatonin levels in rats. Toxicol Ind Health. 2005 Mar;21(1-2):27-31.

Lai, H, Singh, NP, Melatonin and a spin-trap compound block radiofrequency electromagnetic radiation-induced DNA strand breaks in rat brain cells, Bioelectromagnetics, 18, 446-454, 1997a

Oktem F, Ozguner F, Mollaoglu H, Koyu A, Uz E. Oxidative damage in the kidney induced by 900-MHz-emitted mobile phone: protection by melatonin. Arch Med Res. 2005 Jul-Aug;36(4):350-5.

Oral B, Guney M, Ozguner F, et al. Endometrial apoptosis induced by a

900-MHz mobile phone: preventive effects of vitamins E and C. Adv Ther. 2006 Nov-Dec;23(6):957-73.

Ozguner F, Aydin G, Mollaoglu H, Gokalp O, Koyu A, Cesur G. Prevention of mobile phone induced skin tissue changes by melatonin in rat: an experimental study. Toxicol Ind Health. 2004 Sep;20(6-10):133-9.

Ozguner F, Oktem F, Armagan A, et al. Comparative analysis of the protective effects of melatonin and caffeic acid phenethyl ester (CAPE) on mobile phone-induced renal impairment in rat. Mol Cell Biochem. 2005 Aug;276(1-2):31-7.

Ozguner F, Bardak Y, Comlekci S. Protective effects of melatonin and caffeic acid phenethyl ester against retinal oxidative stress in long-term use of mobile phone: a comparative study. Mol Cell Biochem. 2006 Jan;282(1-2):83-8.

Wood, A.W., Armstrong, S.M., Sait, M.L., Devine, L. and Martin, M.J., Changes in human plasma melatonin profiles in response to 50 Hz magnetic field exposure, Journal of Pineal Research, 25, 116-127, 1998

Wood AW, Loughran SP, Stough C. Does evening exposure to mobile phone radiation affect subsequent melatonin production? Int J Radiat Biol. 2006 Feb;82(2):69-76.

Yamaguchi H., Hosokawa K., Soda A., Mizamoto H., Kinouchi Y. (1993): Effects of seven months exposure to a static 0.2 T magnetic field on growth and glycolytic activity of human gingival fibroblasts, Biochem. Biophys. Acta, v. 1156, p. 302-306.

Yariktas M, Doner F, Ozguner F, Gokalp O, Dogrutt H, Delibas N. Nitric oxide level in the nasal and sinus mucosa after exposure to electromagnetic field. Otolaryngol Head Neck Surg. 2005 May;132(5):713-6.

ABSTRACTS

Dr. Henry Lai, a well-known bioelectromagnetics researcher at the University of Washington, Seattle, has compiled a 97-page collection of abstracts from studies conducted between 1995 and 2000. The list, in pdf format, can be found on the Research page of the EMR Network's web site. As the web site points out, "80% of these studies demonstrate some kind of biological effect."

REPORTS

Potential and Actual Adverse Effects of Radiofrequency and Microwave Radiation at Levels Near and Below 2 uW/cm^2, is a 200-page report by Dr. Neil Cherry, of Lincoln University, New Zealand.

7. HOW DO WE KEEP OURSELVES AND OUR FAMILIES SAFE FROM RADIATION?

Invisible electromagnetic radiation (EMR) emanating from diverse sources which include power lines, home wiring, computers, televisions, microwave ovens, photocopy machines, and cell phones surrounds us all the time.

An increasing number of scientists are concerned that EMR poses serious health effects ranging from childhood leukemia to brain tumors. Scientists are continuing to unravel the precise health dangers of EMR. While they are doing so, it makes good sense to avoid these potentially dangerous frequencies as much as possible.

> **What you need to know**
>
> The technology to protect against radiation was developed in a large, on-going research project at the Catholic University of America (CUA), Department of Physics in Washington, D.C.
>
> The technology works by blocking the EMF-induced physical-chemical interactions around the cell membrane receptors, probably by preventing cooperative sensing from happening. In this way, the cascade of biochemical events which would result inside the cell as a function of membrane receptors sensing the constant, oscillating EMF, is prevented from being triggered. The technology re-establishes the situation around the cell as if it was not exposed to EMF.
>
> Dr. Igor Smirnov invented an EMR shielding neutralizer. This device was designed to compensate for the negative effects of EMR on the human body. This device could be used to protect you from your handphone radiation.
>
> Dr Igor Smirnov invented the WaveRider. This powered device produces specific

> frequencies that counter adverse effects associated with our exposure to the electromagnetic radiation. This device generates a field large enough to protect a small house.
>
> Laboratory studies showed the WaveRider technology could reduce SAR values in 80% of experimental points inside the "phantom head".
>
> Laboratory studies also demonstrated that the WaveRider technology partially relieved the effects of mobile phone irradiation on human astrocyte cells.

Man-made electromagnetic fields (EMF) are structurally different from the EMFs found in nature, including the subtle low frequency electromagnetic fields that our body's cells use to communicate with each other.

As shown in the figures below, the EMFs in nature are random (or "noisy") in their frequency, amplitude and waveform whereas artificial EMFs are regular and constant in their frequency, amplitude and waveform. For example, EMF from cell phones appear as regular bursts of microwave radiation. EMF from computer monitors are saw-tooth shaped. EMF from AC-powered devices are sinusoidal. Our cells respond negatively to such regular EMFs, become stressed and try to protect themselves. Below is a summary setting out our cells response to man made EMFs.

Further research has now concluded that cell phones initiate cellular and DNA chaos. They do this through a variety of initially unanticipated non-thermal, biological effects. Information piggybacking on the radio frequencies emitted and received from antenna of cell phones is called an information carrying radio wave (ICRW).

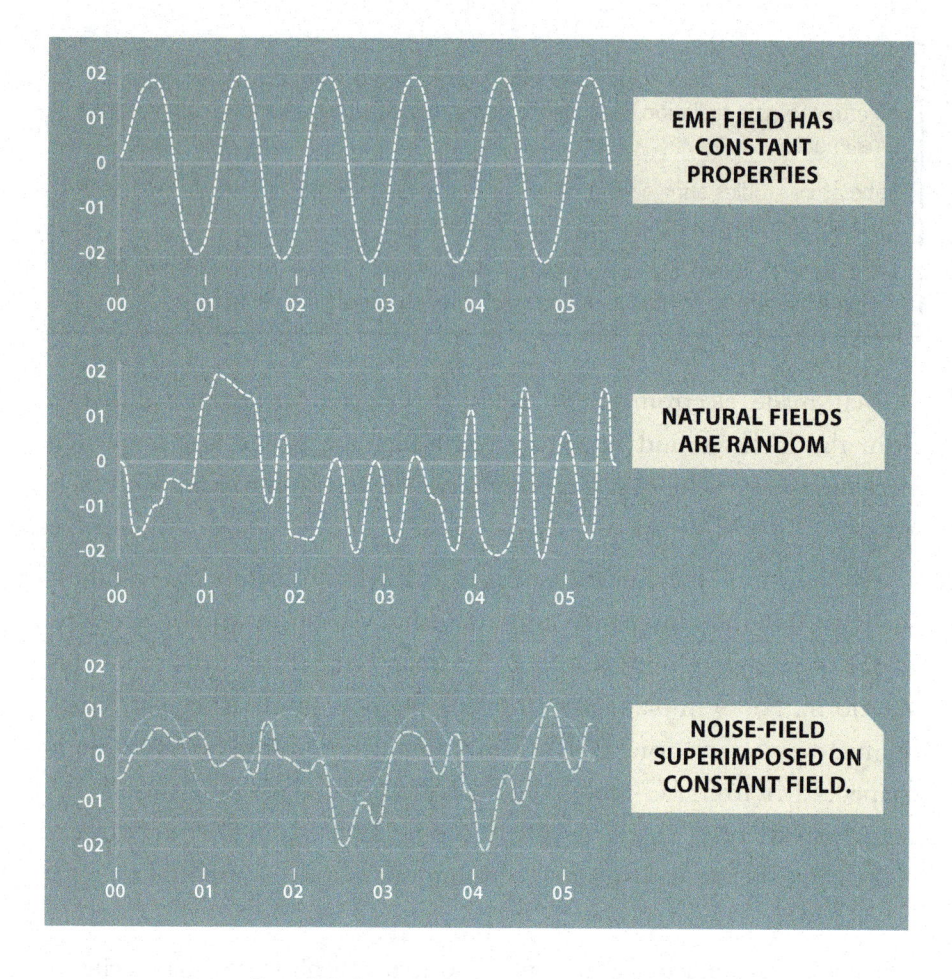

Fig. 16: Differences in Electromagnetic Fields.

It is a frequency that conveys specific packets of information, which allows for the transmission of various features of cell phones, i.e. voice, text or graphics, etc. Emerging science has discovered that IRCW creates a negative domino effect

---

Here's what happens when your cell is exposed to cell phone radiation:

- A sensor in the membrane of the biological cell is triggered by the invading radio wave.

- The sensor is triggered by changes in the molecular nature of intracellular water that has become "chaotic".

> - Once the membrane recognition occurs, a series of protective biochemical reactions are initiated inside the cell as a means of cellular protection.
> - Among these protective reactions are stress protein responses that serve to effectively "harden" the cell membrane and disrupt active transport.
> - The "membrane hardening" effect then causes an intracellular build-up of waste products and toxins, including highly reactive free radicals.
> - These free radicals have been shown in studies to result in cellular dysfunction (evidenced by studies showing leakage in the blood-brain barrier following EMR exposure). They also interfere with normal DNA repair processes (evidenced by studies showing the presence of micronuclei in cells following EMR exposure). These effects have been shown to lead to formation of disease. Several experiments have shown these effects eliminated when EMR exposure was removed.

This domino effect on human cells is the problem. ICRW is a frequency that had never before existed in nature. Thus, it is perceived as foreign and toxic to our cells. Our body as explained below tries to protect itself from ICRW.

This attempt to protect itself against ICRW results in an energy-depleted, weakened immune system. This in turn makes the immune system more vulnerable to environmental toxicity.

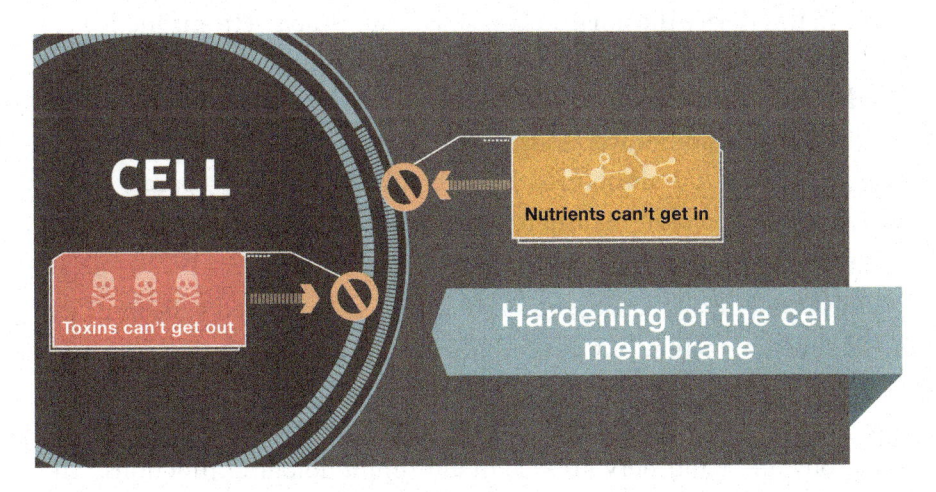

Fig. 17: Impact of artificial radiation on cell wall

Clearly, an effective protective technology must have the ability to shield the cells from the harmful effects of the ICRW. At the same time, the technology must not interfere with the cell phone's signal. The technological devices developed or in the process of being developed can be categorized into 2 main groups.

One focuses on shielding the cells in the human body by neutralizing the effect of electromagnetic radiation on the human body. The other focuses on reducing the effects of electromagnetic radiation on the cells in the human body. It is proposed to examine both these technological devices in this chapter and to assess their efficacy.

THE NOISE- FIELD TECHNOLOGY

The noise-field technology was initially developed as a result of the research funded by US Army in 1986. The Army was concerned about the effects of radiation from communication equipment on key personnel.

The research uncovered the connection between the constancy in EMF wave and the cell's trigger response. They then discovered that by superimposing a random field on the EMF wave, the EMF-induced biological effects are eliminated.

Dr. Igor Smirnov invented an EMR shielding neutralizer. This device was designed to compensate for the negative effects of EMR on the human body. In this device the noise-field is activated by the electromagnetic field itself. In other words, it is generated only in the presence of the source of EMF. This 'passive' noise-field technology is a unique innovation. It requires no outside power source. The Army version used the 'active' principle where the noise-field is generated by a power source. The passive EMR shielding neutralizer (electromagnetic radiation optimum neutralizer) was patented in US in April 2002 .

The WaveRider is a more recently invented a powered version of the EMR shielding neutralizer. It not only has all the benefits and properties of the passive shield but is a quantum leap beyond it, as it incorporates many of Dr Smirnov's research of the best frequencies that promote and boost our brain and nervous system and immunological functions. The WaveRider invention is based on two US Patents: US 8044376 and US 6369399.

RESEARCH DETAILS OF THE WAVERIDER ANTI-RADIATION DEVICE

The WaveRider device is the one of the few commercially available, affordable, and patented noise-field technology in the world that has been scientifically proven to successfully counteract the effects of electromagnetic radiation on the human body.

The **WaveRider** technology works in the following simple, yet fascinating way: The particles within its polymer compound, once stimulated by EMF/EMR, oscillate, and emit a low frequency noise-field that superimposes itself over the harmful ICRW. This unique process creates an incoherent, bio-friendly field, thereby practically negating the otherwise inappropriately triggered protective responses by the body's key systems.

Metaphorically speaking, WaveRider polymer technology acts as a cloaking and deactivating mechanism to the body-foreign ICRW. Since the cloaked ICRW is not considered a foreign invader anymore, the body doesn't feel the need to protect itself, hence no biological disturbance occurs. Simply put, as soon as the WaveRider technology is applied, the otherwise troubling radiation is interpreted as just another vibration that basically has become meaningless and harmless to the body.

WAVERIDER EMR SHIELDING POLYMER APPLICATION

The special polymer compound used in the WaveRider has a high dielectric constant. The EMR shielding polymer material was proven to produce the biological protective effect. This polar polymer material was tested by Underwriters Laboratories and received a UL recognition mark in March 2001.

EMR shielding polymer material helps to avoid the adverse cell response. It "shields" the cellular structures of the body against the harmful effects of EMR. It does this by superimposing an optimal random field or 'noise field' on the bio-effecting electromagnetic field. It neutralizes the negative effects of EMR by changing the quality of the field. In other words the quality of the field is changed from one that can cause cellular damage to a harmless one.

> **Benefits of the WaveRider EMR shielding polymer**
>
> @ In the presence of external electromagnetic radiation (EMR), it emits subtle low frequencies electromagnetic oscillations (the 'noise field').
>
> @ These oscillations resemble resonance frequencies of living cells in the body. These oscillations compensate and neutralize the harmful effects of EMR generated by cell phones and electronic devices.
>
> @ They also provide support for cellular functions in the body. This leaves you feeling healthy and well even when you are exposed to EMR.
>
> @ It also will not disturb the function and operation of your cell phone.

This process makes the resulting field random and the exposure to it neutral on a biological level. In other words, the incident EMR is no longer seen as a threatening signal to the cell. It no longer triggers the cellular defense which causes the damage related to EMR toxicity. This occurs without disturbing the function and operation of your cell phone or wifi signals.

COMPARISON OF THE DIFFERENT EMR PROTECTION DEVICES

In the chart below, different types of technological devices for protecting against EMR (including the Wave Rider) are assessed on the requisite criteria of therapeutic qualities that an EMR shielding device has to satisfy to be effective. The assessments of the different devices are made based on scientific research and documentation available for each of the different devices.

Qualities \ Device	Device A WaveRider	Devices B Earthcalm	Devices C SafeSpace	Devices D Blushield
Protection against Handphone EMR	/	/	/	/
Protection against Wifi EMR	/	/	/	/
Protects against computers and peripherals EMR	/	/	/	/

Protects against Television EMR	/	/	/	/
Household appliances EMR	/	/	/	/
Can offer EMR protection to a radius of 30 feet (9m)	/	X	X	X
Protects one person against EMR	/	/	/	/
Protects many people against EMR	/	/	/	/
Does not Block Handphone Signals	/	?	?	?
Does not disrupt body's natural signaling	/	/	/	?
Phantom Head Experiments Conducted	/	X	X	X
Offers brain cell protection	/	?	?	?
Protects against free radical cellular damage	/	?	?	?
Protects against cellular damage	/	?	?	?
Protects damage against cellular mitochondria	/	X	X	X
Uses "Noise Field Theory" Principles	/	X	X	X
One product that does all the above	/	X	X	X

Legend

/ - The device/s has/have the requisite quality

X - The device/s does/do not have the requisite quality

? - We cannot determine if the device/s has/have the requisite quality

Outlined below are the different devices that the Wave Rider is being compared to.

Device A – WaveRider www.againstradiation.com

Devices B – Earthcalm products www.earthcalm.com

Devices C - EMF protection products www.safespaceprotection.com

Devices D – Powered EMF protection www.blushield-global.com

From the above comparisons, we can clearly see that the The WaveRider is the only device that satisfies all the above therapeutic properties to guard against EMR. The other systems use multiple devices to achieve some of the results. Very often many different devices have to be used to produce some of the results the WaveRider alone produces.

> **Disadvantages of devices that reduce the EMR intensity**
>
> Any type of device that claims to reduce or suppress EMR intensity can create even worse problems for the cellular structures of the body as briefly described below:-
>
> 1. First of all in attempting to reduce the man made EMF, it also will suppress the natural electromagnetic processes in the cells. This is because these processes are thousand times weaker than EMR generated by any electronic appliances. When the natural electromagnetic processes of the body are suppressed or interfered with, you could become sick.
>
> 2. It will create distortion of transmitted signals and worsen the reception of cellular phones, because these devices are based on ferromagnetic materials or high density metals.
>
> Besides these devices only reduce the radiation by 15-20%.

EMR REDUCING DEVICES

Taking into consideration that most of the appliances (cellular phones, computers, etc.) are usually located in a very close proximity to the human body, it is reasonable to admit that shielding devices which reduce electromagnetic fields will, first of all, suppress and disturb electromagnetic processes in living cells of the human body.

THE WAVERIDER – THE POWERED ANTI-RADIATION PROTECTOR

The WaveRider addresses many disadvantages in current methods for the protection against electromagnetic radiation and provides related benefits.

The WaveRider device includes a housing, a solenoid operably connected to a driver and a polymer. The solenoid generates incident radiation which results in the polymer emitting electromagnetic oscillations at specific frequencies. These specific frequencies counter adverse effects associated with the subject's exposure to the electromagnetic radiation. The solenoid includes a two frequency mode that generates two carrier frequencies of incident radiation. The carrier frequencies are at higher frequencies than the oscillation frequencies.

Carrier frequencies may independently or collectively induce oscillation of the polymer materials. The WaveRider device is powered by a 12V wall DC power supply.

The WaveRider uses a solenoid that is constructed from multiple turns of thin wire and has a current rating of about 300 mA and has a frequency response adequate for operating in the frequency range between about 7.0 Hz and 15.0 Hz.

The WaveRider protects against:

*Remote sources of EMR and is therefore protective against any EMR emitting device. This includes computers, a computer peripherals, a cellular telephones, a personal communication devices, televisions, audio systems. It also includes household appliances that may intentionally and/or unintentionally emit electromagnetic radiation.

*It also protects against EMR emitted from indoor or outdoor power lines. Though the Wave Rider is operable outdoors, it can be used indoors to provide protection within an indoor room, such as within a 30 feet (9 metre) radius. The number of subjects that may be protected with the present device is only limited by the protected area. Thus, a single device can protect a few people within the operating area.

*These emitted protective frequencies prevent or reduce the ordinarily deleterious effect from EMR exposure, such as high frequency EMR, on biological processes. Deleterious effects protected against may include changes in viscosity, pressure or water content of bodily fluids such as interstitial fluid, blood, and the like.

> *These frequencies also protect against negative changes in body pH, oxygen content, hydration, mitochondrial activity, hormone levels and the like. It may also prevent or reduce the accumulation of free radicals in response to exposure to EMR.

Fig. 18: The WaveRider

THE WAVERIDER IMPACTS CALCIUM AND OR SODIUM IONS

The experimental data provide evidence of the peak interference spectra for Calcium and Sodium ion transportation for the following applied frequency "windows" of 7.8 Hz and 15 Hz. The housing is constructed of a plastic material permeable to the appropriate incident frequencies.

The protective features of the WaveRider may act by targeting or affecting Calcium and/or Sodium ions in the user's body by emitting and optionally carrying the protective frequencies from 1 to 60 Hz. The WaveRider is likely to affect localization of Calcium and/or Sodium ions in the body; ion pumps and/or ion channels; chemical or biological reactions involving the interaction, binding or transfer of Calcium and/or Sodium.

Calcium is very important for the function of the organism. Calcium ions contribute to the activity of many enzymes, synaptic transfer, secretion, muscular contraction, proliferation, growth and development by interaction with cells or proteins, such as calmodulin and troponin.

Sodium ions provide for a naturally balanced acid-alkaline medium in the organism and excitation signal transfer processes along the nerve cells. Sodium ions are involved in the function of the "ion pumps" that produce an electrical potential difference across the cell membranes by increasing the density of Sodium in the extracellular medium. The protective effects of the WaveRider may affect the localization of ions such as Sodium or Calcium, activity of ion pumps or ion channels.

The WaveRider has a microcontroller that records the period of time that the solenoid driver circuit is in operation in a `Time of Life` counter. The lifetime of useful operation will vary depending on factors such as the particular polymer used. As a general guidance, it is estimated that the lifetime of useful operation of the WaveRider polymer is approximately 17,500 hours.

RESEARCH ON WATER AND RADIATION

Amplitude modulation consists of encoding information onto a carrier signal by varying the amplitude of the carrier. Amplitude modulation produces a signal with power concentrated at the carrier frequency and in two adjacent sidebands. The lower sideband (LSB) appears at frequencies below the carrier frequency; the upper sideband (USB) appears at frequencies above the carrier frequency.

The sideband power accounts for the variations in the overall amplitude of the signal. From the concept mentioned above it is possible to conclude that Wave-Rider Noise Field Generator low frequency signal superimposed on RF carrier microwave field makes the resulting modulated spectral components of microwave field resemble the characteristics of spatial incoherent field.

Such composite, incoherent low frequency electromagnetic field can affect the hydrogen lattice of the molecular structure of water. It can subsequently modify the electrodynamic properties of water. The resonance interaction, includes both a spatial resonance and a resonance of the oscillating frequency of microscopic orbital currents of protons in water molecular hexagons.

The resonance interaction leads to the process of deviation from the stochiometric composition of water. This lead to the reorganization of water clathrate structures with minimum input of energy. The modification of molecular structure of water can lead to the modification of the electrodynamic characteristics of water such as dielectric permittivity and electrical conductivity [Smirnov 2008].

Below is presented the theoretical concept of electrodynamic processes which lead to the reduction of the rate of absorption of electromagnetic field (SAR) composed of microwave radiation and oscillations of very low frequencies and amplitudes in cell water.

THE ELECTRODYNAMIC MODEL UNDERLYING THE WAVERIDER'S MECHANISM (APPENDIX 3)

Under the influence of applied EMF polar molecules tend to align themselves with the field. Although water has polar molecules, its hydrogen bonding network tends to oppose this alignment. The level to which a substance does this is called its dielectric permittivity.

Dependent on the frequency of applied EMF the dipole may move in time to the field, lag behind it or remain apparently unaffected. The ease of the movement depends on the viscosity and the mobility of the electron clouds. In the wide range of EMF frequencies lower than GHz frequency level (corresponding to microwave thermal effect) the water dipoles move in time to the field.

In the range of extremely low frequency of 0.1 – 1000 Hz (corresponding to the extremely low velocity of movement) the dynamic viscosity of water and the resistance of water dipoles to the alignment (dielectric permittivity) are extremely high (up to 108 at 0.1 Hz) due to hydrogen bonding between molecules (molecular coupling).

In this extremely low range of frequency of applied EMF the water dipoles are able to move in time with low frequency electromagnetic field. As a result they can form multilayer molecular formations which oscillate in accordance with applied low frequency EMF.

In the higher frequency range of kHz to GHz generated by RF phones the reorientation process may be modelled using a "wait and switch" process where the water molecule has to wait until favourable orientation of neighbouring molecules occurs and then the hydrogen bonds switch to new molecules.

This range of frequencies of RF phones is related with the ease of the movement of water dipoles resulting in chaotic Brownian movement of water molecules.

In the process of Brownian movement water molecules located in close proximity to each other develop the "friction effect". That results in an increase of the level of absorption of EMF energy emitted by RF phone and in the generation of heat (called "dielectric loss."). This process creates free radicals in cellular tissue leading to radiation damage.

The low frequency oscillations generated by the WaveRider and superimposed on the carrying frequency of RF phone supports the tendency of water dipoles to align. The aligned water dipoles move in time with low frequency field.

This results in the diploes keeping their normal structures of hydrogen bonding and water molecular formations. It supports the existing proper molecular structuring in water. It also counteracts the tendency of RF to break hydrogen bonding between water molecules and to ease their chaotic movement. This interferes with the process of creating cellular free radicals that creates radiation damage.

SAR test was developed on the basis of multiple measurements of electric potentials inside the water based jelly (the simulated living tissue in "phantom head") exposed to external electromagnetic radiation.

There is a direct correlation between the absorption of non-ionizing electromagnetic radiation by the exposed tissue and the magnitude of the electric component of the field applied to the tissue.

RESEARCH STUDY ON THE WAVERIDER

The RF phone was attached to "phantom head." in standard operation position. The SAR values were calculated based on the measurements of

E-field with the help of FCC certified software program. The series of measurements in 242 points were accomplished for RF phones without the WaveRider and with the WaveRider placed at the distance of 7 feet from the Uni-Phantom head exposed to RF phone radiation. The tested 'phantom head" was filled with head tissue simulating liquid of the following electrical parameters measured before the test at 835 MHz:

Relative Dielectricity 41.04   5%
Conductivity 0.89 mho/m   5%

The brain and muscle simulating mixtures consisted of a viscous gel using hydroxethylcellullose (HEC) gelling agent and saline solution. Preservation with a bactericide was added and visual inspection was made to ensure air bubbles were not trapped during the mixing process. The mixture was calibrated to obtain proper dielectric constant (permittivity) and conductivity of the desired simulated tissue.

The measurements in this investigation were taken to simulate the RF exposure effects under worst-case conditions. Precise laboratory measures were taken to assure repeatability of the tests. The tested RF phone complied with the requirements in respect to all parameters subject to the test.

The influence of the WaveRider signals on the resulting microwave signal of RF phones did not significantly affect the air measurements of RF phone signals and subsequently did not lead to any significant distortion of transmitted RF signals. In each experiment SAR values were measured in 242 points of "phantom head.".

The influence of the signals on RF phones in this experiment did not change the location of "Hot Spots". The "Hot Spots" remained in the same location as without the influence of generator, and their amplitudes decreased in 80% of data points.

In 65% of the data points there was observed a significant decrease of SAR values in the range of 10% to 40%. The placement and function of the WaveRider at the distance of 7 feet from the "phantom head" led to the reduction of the majority of SAR values. 9 SAR values out of 12 meaningful SAR values in this experiment were reduced in the range of 2.1% - 12.6%, and only 1 SAR value increased by 3.5%.

The reduction of SAR values calculated on the basis of E-field probe measurements inside the "phantom head" confirmed that incoherent low frequency oscillations generated by the WaveRider actually increased the value of dielectric permittivity of the simulating brain tissue jelly resulting in the reduction of SAR values in the "phantom head."

The research "R&D Testing SAR Evaluation", Test Report No: R&D 20071102 was conducted at FCC certified RF Exposure Laboratory, Escondido, California. It showed the reduction of SAR values in 80% of experimental points inside the "phantom head". This was for three different models of RF phones functioning at 835 MHz (242 points were measured for each RF phone with and without the WaveRider placed at the distance of 7 feet from "phantom head").

In other words as a result of using the WaveRider technology there was a reduction of SAR values in 80% of experimental points inside the "phantom head".

LABORATORY STUDY ON HUMAN ASTROCYTE CELLS

The results for both experiments reveal that when cells are exposed to mobile phone irradiation, astrocytes growth is significantly decreased within the first 3-4 days. Then astrocytes metabolic activity begins to increase. This most likely due to the adaptation effect. Cells structure and function are constantly modified in response to changing environmental influences.

The negative effect of mobile phone irradiation on astrocytes was partially relieved when the WaveRider was placed 30 ft from the treated plate (7.2% average increase of astrocytes metabolic activity with compared to cell samples without WaveRider influence). This represents a single iteration of this particular experiment. It does support the previously observed negative effect of cell phone irradiation on astrocyte growth.

In other words as a result of using the WaveRider technology the effects of mobile phone irradiation was partially relieved on human astrocyte cells.

WHAT FUTURE ARE WE GOING TO CHOOSE?

As a neurosurgeon, Dr. Vini Khurana is on the frontlines of the brain tumor epidemic. He predicts that between 2008 and 2012, a large number of people will have been using cell phones long enough (10 years or longer) to witness the explosion of brain tumors and other health issues.He also believes that within five years, the scientific evidence will overwhelmingly and irrefutably prove the health disaster from all things wireless.

The WaveRider is a timely necessity for all those who desire to "err on the side of safety." It is the only device that has all the therapeutic properties to guard against EMR and to successfully intervene against a fast growing, electropollution-triggered health epidemic.. The other systems use multiple devices to achieve some of the results. (Please see comparison chart)

Given that there is no scientific evidence that man made EMFs are going to do? We can choose to be safe or wait until the case against radiation toxicity is airtight so that we can be dead right in taking preventive action. So what are we going to do to protect ourselves against the certain dangers of electropollution? What are we going to do to protect our children's health? How are we going to protect the genetic future of your descendants from the hazards of radiation pollution? Our unborn children are waiting to hear our answers.

REFERENCES

Smirnov, I.V., Fisher H.W. and Pisarek S. (2009) "Thermographic Evaluation of the MRET-Shield Polymer on the Reduction of Thermal Effects Caused by Radio Frequency Radiation" Explore Magazine, Vol.18, No.1: 14-17, USA

Smirnov, I.V. (2008) "The Effect of MRET Polymer Compound on SAR Values of RF Phones" Journal of Microwave Power & Electromagnetic Energy, Vol.42, No.1: 42-54, USA

Smirnov, I.V., Fisher H. and Pisarek S. (2008) "The Beneficial Effect of MRET-Shield on Blood Morphology in vitro Following the Exposure to Electromagnetic Radiation of Cell Phone" Explore Magazine, Vol.17, No.4, USA

Smirnov, I.V. (2006) "Polymer Material Providing Compatibility between Technologically Originated EMR and Biological Systems" Explore Magazine, Vol.15, No.4: 26-32, USA

Smirnov I.V. (2005) "Comparative Study of the Effect of Microwave Radiation Neutralizers on Physiological State of Human Subjects" Explore Magazine, Vol.14, No.5: 29-44, USA

Smirnov, I.V. (2002) "Electromagnetic Radiation Optimum Neutralizer" Explore Magazine, Vol.11, No.1: 45-50, USA

Smirnov, I.V. (2001) "Electromagnetic Radiation Optimum Neutralizer" RSA Magazine (Radiation Safety Associates), December, 2001, USA

APPENDIX 1: THRESHOLD LEVELS AND STANDARDS FOR NON-IONISING RADIATION

Manufacturer	Model	SAR Output (W/Kg)
Motorola	V195	1.6
Motorola	Rival	1.59
Sony Ericsson	Satio (Idou)	1.56
Blackberry	Curve 8330	1.54
Nokia	E71x & X6	1.53
LG	Rumor	1.51
Blackberry	Bold	1.51
Samsung	S3650 Corby	0.75
Samsung	SGH-G800	0.23
Samsung	Blue Earth	0.196

SAR is expressed in Watts per kilogram
Current UK Standard = 1.0W/Kg
Current US Standard = 1.6W/Kg

Table A: SAR levels for different hand phone models

San Francisco Govt. has made it mandatory for the industry to display SAR value for each phone. (USA Today 14 July, 2010)

APPENDIX 1: THRESHOLD LEVELS AND STANDARDS FOR NON-IONISING RADIATION

Iphones: Apple Iphone 3G North America

Frequency	Band	IC 1g SAR Limit (W/kg)	Body	Ear
GSM	850	1.6	1.030	0.521
GSM	1900	1.6	0.522	1.290
UMTS	II	11.66	0.402	1.388
UMTS	V	1.6	0.733	0.733
Wi-Fi		1.6	0.088	0.779

Table B: SAR levels for different iPhone models

Source: www.Sarvalues.com

Comparison of Standard Threshold Values and Recommendations (Electromagnetic fields, non ionizing radiation)	Total Power Density
Standards, GSM 1800/GSM 1900/ UMTS / DECTS (e.g)	
FCC / ANSI - USA	10,000 uW/m$_2$
Germany, England, Finland and Japan	10,000 uW/m$_2$
Belgium	1,200 uW/m$_2$
Switzerland and Italy	90,000 uW/m$_2$
Recommendations / References (e.g.)	
Ecolog Study, Germany (ECOLOG 2000)	10,000 uW/m$_2$
Cellular tower radiation - significant exposure level, 95th percentile (this study)	6,300 uW/m$_2$
Salzburg, Austria (RESOLUTION 2000)	1,000 uW/m$_2$
Cellular tower radiation - median level. 50th percentile (this study)	200 uW/m$_2$
High exposure, Oeko-Test (OEKO TEST 2001)	100 uW/m$_2$
EU Parliament (STOA 2011)	100 uW/m$_2$
Cellular tower radiation - background level, 20th percentile (this study)	10 uW/m$_2$
Low Exposure, Oeko-Test (OEKO TEST 2001)	10 uW/m$_2$
Nighttime exposure. Baubiology Standard (SBM 2000)	0.1 uW/m$_2$
Successful Communication with GSM mobile phone, system coverage requirements	0.001 uW/m$_2$
Natural cosmic microwave radiation (MAES 2000)	0.000001 uW/m$_2$

Table C: Comparison of Threshold Values for emf and non-ionizing radiation from Haumann Thomas, et al, Germany

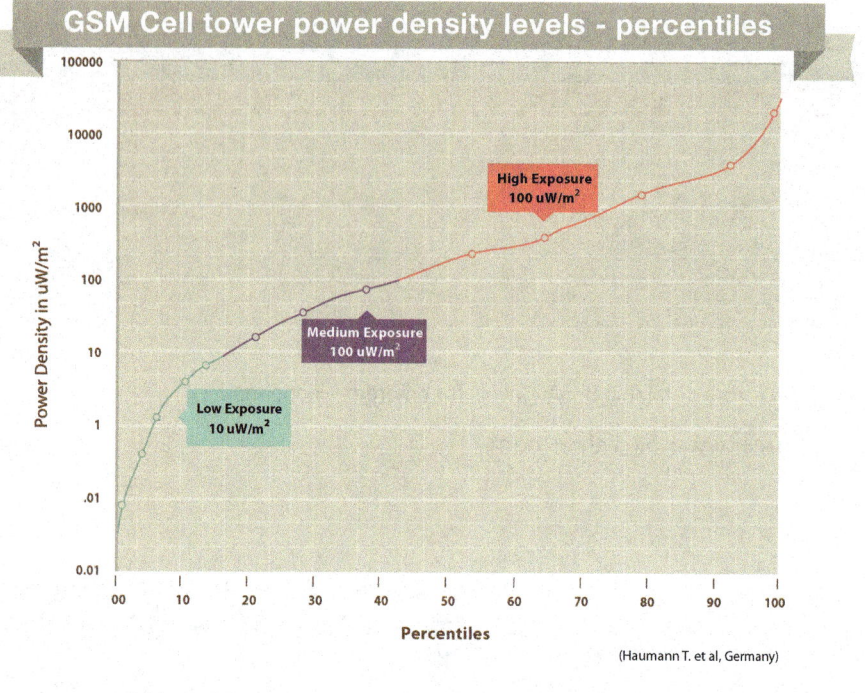

Table D: GSM cell tower power density levels and exposure

- Building Biology Institute, Germany, provided following guidelines for exposure:

a. <0.1 µW/m2 (0.00001 µW/cm2) - no concerns
b. 0.1 - 10 µW/m2 (0.00001 to 0.001 µW/cm2) - slight concern
c. 10 - 1000 µW/m2 (0.001 to 0.1 µW/cm2) - severe concern
d. > 1000 µW/m2 (> 0.1 µW/cm2) - extreme concern

- We recommend safe power limit up to 50 µW/m² with upper limit as 100 µW/m².

India adopts ICNIRP guideline for Power density (Pd) = Frequency / 200, frequency is in MHz.

APPENDIX 1: THRESHOLD LEVELS AND STANDARDS FOR NON-IONISING RADIATION

For GSM900 (935-960 MHz), Pd = 4.7W/m² and
GSM1800 (1810-1880 MHz), Pd = 9.2W/m².

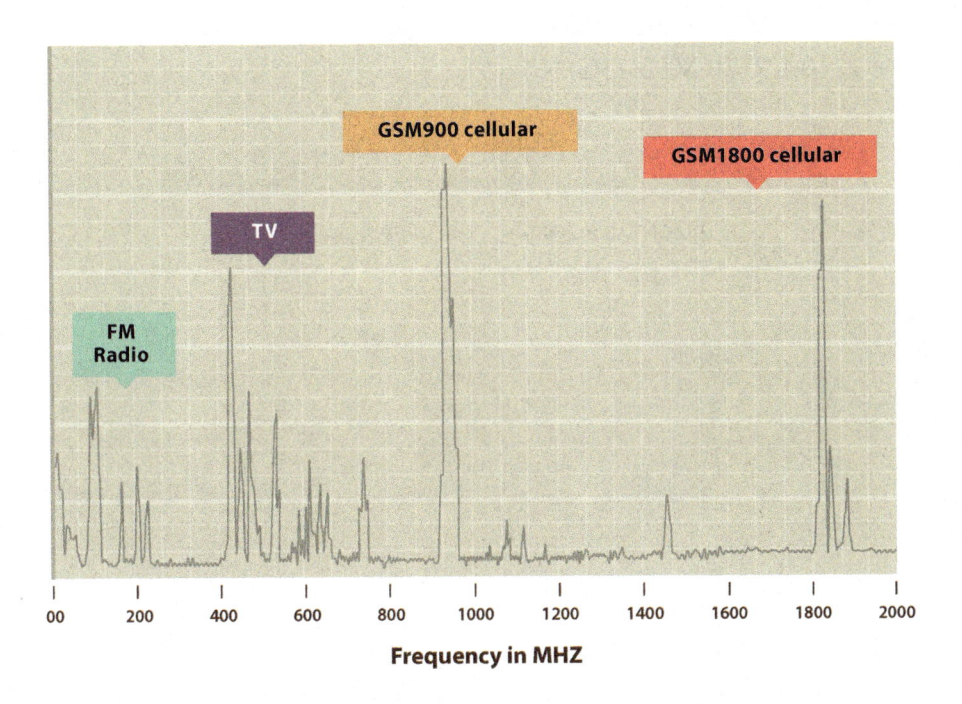

Table E: RF Spectrum near cell tower

H(m)	R(m)	Power density (mW/cm²)
230	803	0.00147
200	806	0.9914.5
150	814	0.00139
100	825	0.00132
50	838	0.00124
0	854	0.00115

Table F: TV Tower Radiated Power Density at Different Height at a distance of 800 m

Antennas on Cell tower transmit in the frequency range of:
- 869 - 890 MHz (CDMA)
- 935 - 960 MHz (GSM900)
- 1805 – 1880 MHz (GSM1800)
- 2110 – 2170 MHz (3G)*

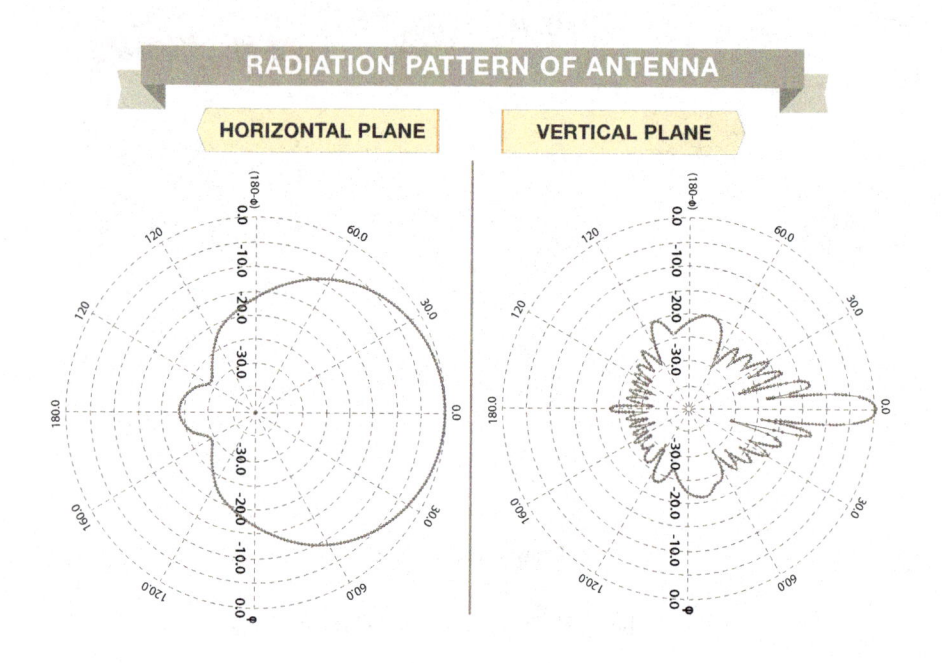

Fig.A: Radiation Pattern of Antenna

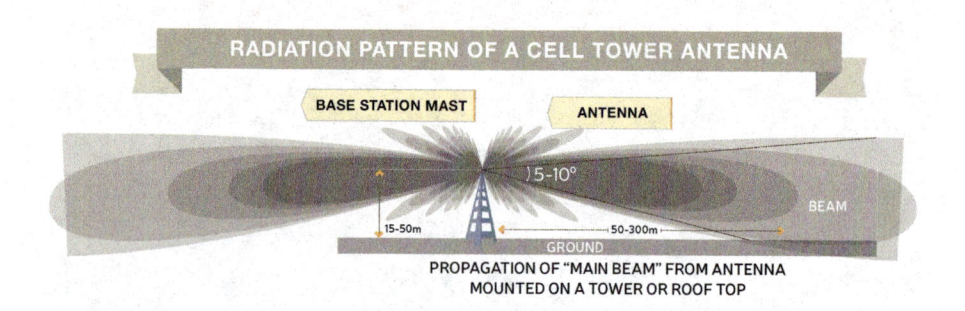

Fig. B: People living within 50 to 300 meter radius are in the high radiation zone (dark blue) and are more prone to ill-effects of electromagnetic radiation

APPENDIX 2: BASICS OF RADIATION THEORY

To understand the fields it is necessary to begin with the electric charge. An atom consists of a nucleus and electrons that are in its orbit. Each electron possesses, a so-called negative charge and the nucleus is made up of protons and neutrons. Protons are positively charged whereas neutrons have no net charge. Each atom contains an equal number of electrons and protons, but their polarity is different and in nature they appear as multiples of equal elementary charge.

Only the electrons in the outer orbit may be dislodged when an external force acts on them. These electrons, when conductors are concerned, are loosely bound, so we may say that conductors have so-called free electrons, while insulators require a strong force which would make the electrons leave its atom.

THE MAGNETIC FIELD

Around a permanent magnet there is something that make an iron particle jump through space to the magnet. Obviously, there is an invisible entity, a phenomenon that exerts a force on the iron, but as to just what it consists of, nobody knows; the phenomenon is called a magnetic field. A different, but analogous phenomenon extends outward from electrically charged objects, exerting a force on other charged objects; this is called an electric field.

Magnetic fields arise from the motion of electric charges, i.e. a current. They govern the motion of moving charges. Their strength is measured in units of ampere per metre, (A/m) but is usually expressed in terms of the corresponding magnetic induction measured in units of tesla, (T), millitesla (mT) or microtesla (mT).

In some countries another unit called the gauss, (G), is commonly used for measuring magnetic induction (10,000 G = 1 T, 1 G = 100 mT, 1 mT = 10 G, 1 mT = 10 mG). Any device connected to an electrical outlet, when the device is switched on and a current is flowing, will have an associated magnetic field that is proportional to the current drawn from the source to which it is connected. Magnetic fields are strongest close to the device and diminish with distance. They are not shielded by most common materials, and pass easily through them. You have no protection against magnetic fields.

MAGNETOSTATIC FIELDS

Dipoles and induced currents polarized in this way have got their magnetic field which is superimposed to the outer one. The sum of both poles in the first medium gives a reflected wave, and in the other tube is a refracted one (wave reflection and refraction). As a consequence a tension is created in the other medium as well on the very border between the two media. Compared to the electromagnetic field, magnetostatic field passes through living organisms without difficulty. In so doing it produces magnetic polarization in the organism whose effects may be neglected and acts with a force only on ions which are in motion.

ELECTROSTATIC FIELDS

When electrostatic field reaches the human organism, it induces the separation of positive and negative charge. At the spot where it has entered, the surface is negatively charged and on the spot where it leaves, there is positive charge. Since the organism is not a perfect conductor it is difficult to say that there is no electrostatic field even after that short separation of charge. As the electrostatic field is a potential field it may disturb the electric potential and the normal functioning of the organism.

BASIC LAWS OF ELECTRICITY AND MAGNETISM

It is a basic law of physics as electricity flows, a magnetic field is created in the surrounding space. In 1820, Hans Christian Oersted accidentally discovered that passing a current through a wire would cause nearby compass needles to rotate. Electricity can give rise to magnetism!

The circulatory system is an excellent conductor of electricity. According to Ampére's Law, the flow of electricity through the blood vessels and other fluid compartments in the body must create a magnetic field in the space around the body.

Essentially electricity gives rise to magnetism and magnetism gives rise to electricity. So in biological systems, bioelectricity gives rise to biomagnetism and biomagnetism gives rise to bioelectricity

THE ELECTROMAGNETIC SPECTRUM

Electromagnetic fields (EMF) consist of electric (E) and magnetic (H) waves travelling together. They travel at the speed of light and are characterised by a frequency and a wavelength. The frequency is simply the number of oscillations in the wave per unit time, measured in units of hertz (1 Hz = 1 cycle per second), and the wavelength is the distance travelled by the wave in one oscillation (or cycle).

EMFs are sometimes called radiation when the frequency is measured in kilohertz and above. EMFs are categorized, according to their frequency or wavelength, in the electromagnetic spectrum. The spectrum spans an enormous range of frequencies.

The lowest frequency EMFs (below 3000 Hz, or 3 kHz) are called extremely low frequency (ELF) fields. They are mainly generated by AC current devices and power lines and usually have a frequency of 60 Hz (North America) or 50 Hz (elsewhere).

Frequencies in the kHz (thousand hertz) and low MHz (megahertz, million hertz) region are called radio frequency (RF) fields or radiation and are used for radio and TV broadcasting and two-way radio systems.

The RF region is – arbitrarily – broken up into a further alphabet of frequencies like EHF, SHF, UHF, VHF, HF, MF, LF and VLF. In addition to radio transmitter equipment, also computer displays radiate RF waves in the kHz region in addition to the ELF fields associated with their AC power supply.

<u>Frequencies in the high MHz and the GHz (gigahertz, billion hertz) region are called microwave (MW) fields or radiation and are used for cell phones, personal communication systems, microwave ovens and radar systems.</u>

Frequencies above microwave and below visible light ($10^{12} - 10^{14}$ Hz) are called infrared radiation. This type of EMF is radiant heat emitted from hot objects like ovens.

Visible light is a narrow band of frequencies around 10^{15} Hz. Visible light is emitted from atoms when electrons in their outer shells change orbits around the nucleus of the atom.

Frequencies in the spectral region above visible light are - according to increasing frequencies - ultraviolet, X-rays, gamma rays and nuclear radiation. These types of radiation are categorized as ionizing radiation, whereas all frequencies below ultraviolet are called non-ionizing radiation.

OTHER SOURCES OF RADIATION

Ionizing radiation is radiation that has sufficient energy to remove electrons from atoms. One source of radiation is the nuclei of unstable atoms. For these radioactive atoms (also referred to as radionuclides or radioisotopes) to become more stable, the nuclei eject or emit subatomic particles and high-energy photons (gamma rays).

This process is called radioactive decay. Unstable isotopes of radium, radon, uranium, and thorium, for example, exist naturally. Others are continually being made naturally or by human activities such as the splitting of atoms in a nuclear reactor. Either way, they release ionizing radiation. The major types of radiation emitted as a result of spontaneous decay are alpha and beta particles, and gamma rays. X rays, another major type of radiation, arise from processes outside of the nucleus.

IONIZING VERSUS NON-IONIZING RADIATION

Ionizing radiation contains so much energy in its individual quanta of energy - photons - that it is able to knock out electrons from their orbits in the atom shells. This creates free radicals in living matter. It also increases the risk of chromosomal damage, fetal abnormalities and cancer. These health consequences of ionizing radiation were disputed since the beginning of the 20th century. They became generally accepted by the middle of the century.

Radiation causes ionizations in the molecules of living cells. These ionizations result in the removal of electrons from the atoms, forming ions or charged atoms. The ions formed then can go on to react with other atoms in the cell, causing damage. An example of this would be if a gamma ray passes through a cell, the water molecules near the DNA might be ionized. The ions might react with the DNA causing it to break.

Non-ionizing radiation contains too little individual photon energy to knock out electrons from their orbits in the atoms. If non-ionizing radiation has health effects, there is some other mechanism in play. However, scientific studies have shown that non-ionizing radiation indirectly may lead to increases in the level of free radicals in tissue. This would potentially lead to the same consequences as ionizing radiation

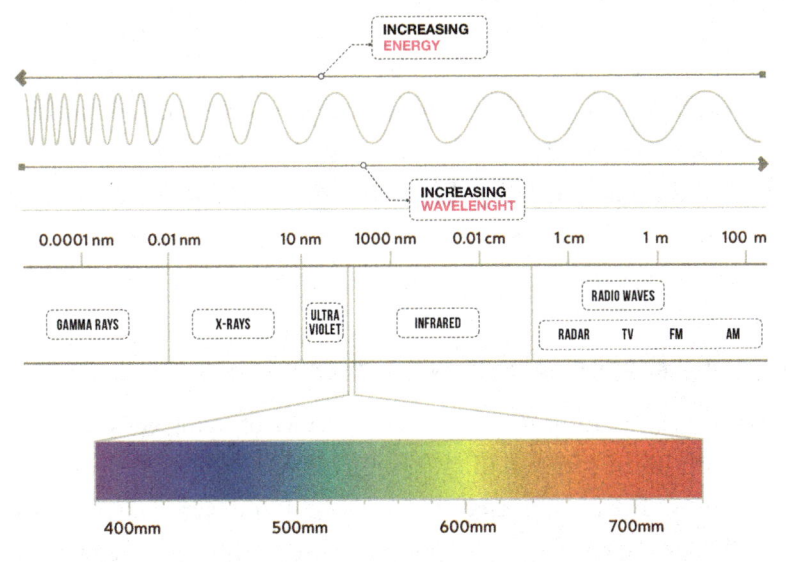

Fig.1: The Electro Magnetic Spectrum

Research has revealed that there are significant biological effects of low-level (low intensity) non-ionizing radiation. These effects are not related to heating, the so-called athermal effects. The mechanisms responsible for athermal effects of low-intensity radiation or fields are just beginning to be understood by scientists. There is an intensive discussion amongst scientists over the potential health effects.

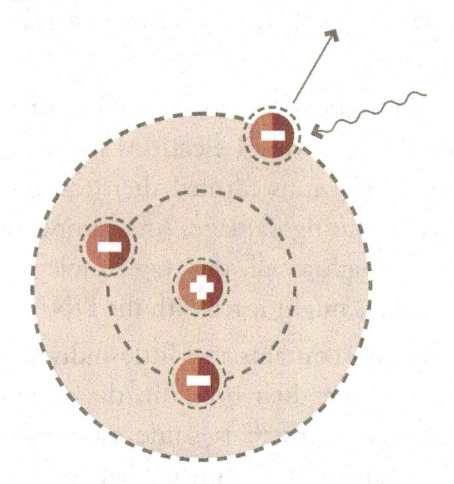

Fig. 2: An atom with an electron being knocked out by an ionizing electromagnetic wave

Energy, usually the electromagnetic form, or that emitted by the sun is called radiation or solar radiation. Energy is transmitted, or travels, through space in different wave lengths, strengths, or intensities of electromagnetic fields (EMF). Electromagnetic energy can come from natural sources, like those that were built-up during thunderstorm formations and dispersed throughout the atmosphere.

WHAT IS THE EFFECT OF IONIZING RADIATION?

Radiation causes ionizations in the molecules of living cells. These ionizations result in the removal of electrons from the atoms, forming ions or charged atoms. The ions formed then can go on to react with other atoms in the cell, causing damage. An example of this would be if a gamma ray passes through a cell, the water molecules near the DNA might be ionized and the ions might react with the DNA causing it to break.

At low doses, such as what we receive every day from background radiation, the cells repair the damage rapidly. At higher doses (up to 100 rem), the cells might not be able to repair the damage, and the cells may either be changed permanently or die. Most cells that die are of little consequence, the body can just replace them.

Cells changed permanently may go on to produce abnormal cells when they divide. In the right circumstance, these cells may become cancerous. This is the origin of our increased risk in cancer, as a result of radiation exposure. At even higher doses, the cells cannot be replaced fast enough and tissues fail to function.

An example of this would be "radiation sickness." This is a condition that results after high doses to the whole body (>100 rem). This can occur from a radiation exposure from a nuclear reactor accident, exposure to the after effects of a nuclear bomb. The intestinal lining thereafter is damaged to the point that it cannot perform its functions of intake of water and nutrients, and protecting the body against infection.

This leads to nausea, diarrhea and general weakness. With higher whole body doses (>300 rem), the body's immune system is damaged and cannot fight off infection and disease. At whole body doses near 400 rem, if no medical attention is given, about 50% of the people are expected to die within 60 days of the exposure, due mostly from infections.

If someone receives a whole body dose more than 1,000 rem, they will suffer vascular damage of vital blood providing systems for nervous tissue, such as the brain. It is likely at doses this high, 100% of the people will die, from a combination of all the reasons associated with lower doses and the vascular damage.

There a large difference between whole body dose, and doses to only part of the body. Most cases we will consider will be for doses to the whole body. What needs to be remembered is that very few people have ever received doses more than 200 rem. With the current safety measures in place, it is not expected that anyone will receive greater than 5 rem in one year. Radiation risk estimates, therefore, are based on the increased rates of cancer, not on death directly from the radiation.

Non-Ionizing radiation does not cause damage the same way that ionizing radiation does. It tends to cause chemical changes (UV) or heating (Visible light, Microwaves) and other molecular changes (EMF).

HOW ARE HUMANS AFFECTED BY IONIZING RADIATION?

The effects of a given dose of ionizing radiation on humans can be separated into two broad categories: Acute and Long-Term effects.

Although a dose of just 25 rems causes some detectable changes in blood, doses to near 100 rems usually have no immediate harmful effects. Doses above 100 rems cause the first signs of radiation sickness including:

nausea

vomiting

headache

some loss of white blood cells

Doses of 300 rems or more cause temporary hair loss. There is also more significant internal harm, including damage to nerve cells and the cells that line the digestive tract. Severe loss of white blood cells, which are the body's main defense against infection, makes radiation victims highly vulnerable to disease.

Radiation also reduces production of blood platelets, which aid blood clotting. So victims of radiation sickness are also vulnerable to hemorrhaging. Half of all people exposed to 450 rems die, and doses of 800 rems or more are always fatal. Besides the symptoms mentioned above, these people also suffer from fever and diarrhea. As of yet, there is no effective treatment—so death occurs within two to fourteen days.

In time, for survivors, diseases such as leukemia (cancer of the blood), lung cancer, thyroid cancer, breast cancer, and cancers of other organs can appear due to the radiation received.

HOW IONIZATIONS AFFECT CELLS

Radiation-induced ionizations may act directly on the cellular component molecules or indirectly on water molecules, causing water-derived radicals. Radicals react with nearby molecules in a very short time, resulting in breakage of chemical bonds or oxidation (addition of oxygen atoms) of the affected molecules. The major effect in cells is DNA breaks. Since DNA consists of a pair of complementary double strands, breaks of either a single strand or both strands can occur.

However, the latter is believed to be much more important biologically. Most single-strand breaks can be repaired normally thanks to the double-stranded nature of the DNA molecule. The two strands complement each other, so that an intact strand can serve as a template for repair of its damaged, opposite strand. In the case of double-strand breaks, however, repair is more difficult and erroneous rejoining of broken ends may occur. These so-called misrepairs result in induction of mutations, chromosome aberrations, or cell death.

CHARACTERISTICS OF DNA DAMAGE BY RADIATION EXPOSURE

Deletion of DNA segments is the predominant form of radiation damage in cells that survive irradiation. It may be caused by

(1) misrepair of two separate double-strand breaks in a DNA molecule with joining of the two outer ends and loss of the fragment between the breaks or

(2) the process of cleaning (enzyme digestion of nucleotides--the component molecules of DNA) of the broken ends before rejoining to repair one double-strand break.

BIOLOGICAL EFFECTS DIFFER BY TYPE OF RADIATION

Radiations differ not only by their constituents (electrons, protons, neutrons, etc.) but also by their energy. Radiations that cause dense ionization along their track (such as neutrons) are called high-linear-energy-transfer (high-LET) radiation, a physical parameter to describe average energy released per unit length of the track. (See the accompanying figure.) Low-LET radiations produce ionizations only sparsely along their track and, hence, almost homogeneously within a cell.

Radiation dose is the amount of energy per unit of biological material (e.g., number of ionizations per cell). Thus, high-LET radiations are more destructive to biological material than low-LET radiations, such as X and gamma rays. This is because at the same dose, the low-LET radiations induce the same number of radicals more sparsely within a cell, whereas the high-LET radiations, such as neutrons and alpha particles,-transfer most of their energy to a small region of the cell. The localized DNA damage caused by dense ionizations from high-LET radiations is more difficult to repair than the diffuse DNA damage caused by the sparse ionizations from low-LET radiations.

ACUTE EFFECTS OF RADIATION

The acute, or more immediately-seen effects of radiation can affect the performance of astronauts. These effects include skin-reddening, vomiting/nausea and dehydration. Other tissue and organ effects are possible. Another term: Acute Radiation Syndrome.

LONG TERM EFFECTS RADIATION

Given that only moderate doses of radiation are encountered (and thus acute effects are not seen) the long-term effects of radiation become the most important to consider. The passage of an energetic charged particle through a cell produces a region of dense ionization along its track.

The ionization of water and other cell components can damage DNA molecules near the particle path but a "direct" effect is breaks in DNA strands. Single strand breaks (SSB) are quite common and Double Strand Breaks (DSB) are less common but both can be repaired by built-in cell mechanisms. "Clustered" DNA damage, areas where both SSB and DSB occur can lead to cell death.

Although "endogenous processes" can lead to DSB, its occurrence due to ionizing radiation (especially the high LET radiation found in space) is an important component of long-term risk . For most cell types, the death of a single cell is no big deal -- cells continually die and are replaced by normal processes.

A more dangerous event may be the non-lethal change of DNA molecules which may lead to cell proliferation, a form of cancer. Research topic: The RBE of alpha particles on stem cells. These single and double strand breaks, or lesions, can be studied with the scanning tunneling microscope.

SOURCES OF IONIZING RADIATION TOXICITY

The sources of radiation pollution involve any process that emanates radiation in the environment. While there are many causes of radiation toxicity (including research and medical procedures and wastes, nuclear power plants, TVs, computers, radio waves, cell-phones, etc.), the most common ones that can pose moderate to serious health risks include:

Nuclear waste handling and disposal – may generate low to medium radiation over long period of times. The radioactivity may contaminate and propagate through air, water, and soil as well. Thus, their effects may not be easily distinguishable and are hard to predict. Additional, some nuclear waste location may not be identified. The main issue with the radiation waste is the fact that it cannot be degraded or treated chemically or biologically.

Mining of radioactive ores (such as uranium ores) – involve the crushing and processing of radioactive ores and generate radioactive by-products. Mining of other ores may also generate radioactive wastes (such as mining of phosphate ores).

Nuclear accidents – an already classic example of such accident is the nuclear explosion at a former Soviet nuclear power plant from Chernobyl that occurred in the mid 1986. Its effects are still seen today. Another example is the 1979 explosion at Three Mile Island nuclear-power generating plant near Harrisburg, PA. The general problems at nuclear weapons reactors are other examples of this type of sources of radiation pollution. Even accidents from handling medical nuclear materials/wastes could have radiation health effects on workers.

Thus, the only options are to contain the waste by storing it in tightly closed containers shielded with radiation-protective materials (such as

Pb) or, if containing is not possible, to dilute it. The waste may also be contained by storage in remote areas with little or no life (such as remote caves or abandoned salt mines).

However, in time, the shields (natural or artificial) may be damaged. Additionally, the past waste disposal practices may not have used appropriate measures to isolate the radiation. Thus, such areas need to be carefully identified and access restrictions promptly imposed.

APPENDIX 3: CELLULAR WATER AND RADIATION

MESENCHYMAL MATRIX

The matrix is the substance between the cells. It is described as having sieve-like properties. It is a semi-fluid, liquid crystal medium, which occupies the space between the muscles, organs and cells of the body. Bio-energy signals traverse the matrix, and the term "matrix" is often used interchangeably with the term 'ground regulation system'. This structure is primarily filled with structured cellular water that is significantly affected by man-made radiation.

It is the space through which food, water and oxygen, must pass through before they diffuse into the cells. It is the space through which, toxins from the cells must diffuse out to before they are hopefully eliminated from the body.

The matrix contains soluble and insoluble fibrins, which provide tensile strength in muscles and ligaments. It also contains adhesive proteins, and space filling proteoglycans. The adhesive proteins include laminin, fibronectin and elastin. These bind together the other structural components of the matrix, and connect the matrix to the cell membrane surface receptors via integrins.

The proteoglycans, not only act as packaging material, they also account for the dielectric and other physiological properties of the matrix. For example, it is these properties, which enable bio-information transport, and selective excretion and resorbtion of urea and electrolytes in the kidneys.

Due to their polar nature and negative charges, proteoglycans bind with many water molecules to form the hydrated colloidal liquid crystalline gel that is responsible for the dielectric properties of the matrix.

The surface of the cell membrane, which has excellent dielectric properties, is called the glycocalyx. Each cell of the body is a powerhouse of swirling electro-magnetic energy. It is the influence of electro-magnetic signals from the mind, brain and autonomic nervous system. Resonating chemical messengers mediate these electromagnetic signals, which are non-material or material information bearing radio wave like signals.

The chemical messengers (hormones) are made up of chains of amino acids or their resonances. The chemical messengers vibrate at coherent frequencies and carry poorly understood encoded information signals. These biological information signals traverse the matrix rapidly.

Often the signals travel much faster than nerve conduction, at the speed of light. A healthy matrix contains water, electrolytes and soluble fibrin, and has excellent electro-conductive and dielectric properties. It is therefore a good electrical conductor of bio-energy signals.

The living matrix has piezoelectric properties so that every movement, tension, pressure, etc. within the connective tissue generates a variety of bio-communication signals. Interference with normal movement and abnormal pressure (blockage), etc., produce abnormal "tone" and aberrant messages within the system as a whole, including to within the cell and to the nuclear level.

The signals work by stimulating receptors at end organ cell membranes and in cell nuclei into physiological activity. They may start with a thought (a coherently resonating information signal itself), or with an information signal from the body's own internal regulatory controls, (e.g. autonomic nervous system activity to do with appetite control or with temperature regulation).

The signals trigger intracellular enzyme systems into various activities, such as making digestive enzymes, or mediating voluntary muscle movement. They may also cause an output of mind-influencing neurotransmitters such as ACTH and beta-endorphin from the anterior pituitary gland, or mood enhancing serotonin from the brain.

GROUND REGULATION SYSTEM

The system of ground regulation has four main systems of communication. One is chemical, as in the many processes described above. The others are electrical impulses through the nerves, electrochemical synapses (found between fibroblasts and between functional organ cells helping them to act together), and electromagnetic vibrations. Thus the wide variety and vast quantity of internal and external information is coded and exchanged in only these four ways.

NEUROTRANSMITTERS AND THE MATRIX

The hypothalamus directly and indirectly controls the release of neurotransmitters and related chemicals called neuropeptides. The hypothalamus is the area of the brain where emotions are generated. Via the hypothalamus, emotional states cause the release of a great variety of neuropeptides including adrenaline, serotonin, dopamine, endorphin, insulin and glutamine. These directly affect the ground substance and thereby the health of the whole body

ELECTROMAGNETIC FIELDS AND THE MATRIX

Many electromagnetic fields originate within the body. All chemical reactions generate electromagnetic fields. Some of the most important electromagnetic fields are generated in the ground substance and in the related connective tissues. Collagen fibers are piezoelectric and pyroelectric. This means that stretched or warmed, they produce an electrical potential. The collagen fibers and the sugar-protein complexes that are attached to them bind water and electrically charged metal ions in the ground substance. Man-made radiation can interfere with these naturally occurring electromagnetic fields by altering the arrangement of water molecules

PIEZOELECTRIC EFFECT AND THE MATRIX

The combination of collagen fibers and sugar-protein complexes produces its highest piezoelectric energy values at 37° C, the temperature of the human body. And the liquid crystal molecular structure of water is highly ordered with minimum energy also at 37°.

This combination allows signaling to traverse the watery medium of the ground substance of the human body easily and with practically no energy loss and no consequent warming (Heine, 1997, p. 142). Thus a disturbance anywhere in the body is registered almost immediately everywhere within the body.

ACUPUNCTURE POINTS AND THE MATRIX

Acupuncture points have a functional connection with the system of ground regulation. The concept of the acupuncture "point" once hindered research greatly. G. Kellner, under the direction of Pischinger at the University of Vienna, searched for special nerve endings in the skin that might be the physical aspects of acupuncture points.

He came to the conclusion that there is no definite physical structure corresponding to acupuncture points (Kellner, 1979). Both the water and chemical contents of this bundle of blood vessels and nerves are electrically conductive. In comparison, fascia tissues are electrically resistant.

BIOPHOTONS

Biophotons are characterized by an extremely high degree of order and can be described as a type of biological laser light which is capable of interference and appears to be responsible for many effects which ordinary incoherent light could not achieve. Its high coherency lends the biophoton wave the capability of creating order and transmitting information while chaotic, incoherent light simply transmit energy.

NEUROPEPTIDES AND NEUROTRANSMITTERS

Dr Candace Pert, a researcher, pharmacologist and professor at Georgetown University, has for the past twenty years studied the movement of neuropeptides in the body. She has shown that "all emotions are healthy, because emotions are what unite the mind and the body". "To repress these emotions and not let them flow freely is to set up a dissonance in the system."

ALBERT SZENT-GYORGI: ELECTRONIC CONDUCTION

Albert Szent-Gyorgi received the Nobel Prize in Chemistry in 1937 for discovery of Vitamin C. He propounded the idea of electronic conduction. According to Albert Szent-Gyorgi, if a number of molecules can be arranged with regularity in close proximity as for example in a crystalline lattice, single electrons cease to belong to one or two atoms only and belong instead to the whole system. A great number of molecules may join to form energy continua, along which energy namely excited electrons may travel a certain distance.

Virtually all of the molecules forming the living matrix are semiconductors. Molecules do not have to touch each other to interact. Energy can flow through the electromagnetic field. The electromagnetic field, along with water, forms the matrix of water. Water can form structures that transmit energy. These coherent water structures can also be disrupted by man-made radiation and initiate cellular damage.

HERBERT FROHLICH: BIOLOGICAL COHERENCE

Dr Frohlich's research goes on to report that coherent vibrations recognize no boundaries, at the surface of a molecule cell, or organism – they are collective or cooperative properties of the entire being. Research on electrically polarized molecule arrays reveals that interactions repeated by the millions of molecules within a cell membrane, tendon, muscle, bone, nerve cell or other structure, give rise to huge coherent or laser like vibrations. Crystalline components of the living matrix act as coherent molecular antennas, radiating and receiving signals.

IONIC DISASSOCIATION IN THE CELLS

August Arrhenius made a fundamental discovery that laid the basis for electrophysiology. This discovery explained how electricity could be conducted throughout the body. It provided the basis for modern theories of nerve conduction, muscle contraction, secretion, and many other processes. He won the Nobel Prize in Chemistry in 1903.

The fundamental discovery was that sodium chloride ions disassociated in solutions, providing the ionic charges necessary for electrical conductivity.

$$NaCl \rightarrow Na^+ + Cl^-$$

We now know that the ionic currents set up by the heart, brain, muscles, the retina, and by other tissues and organs give rise to electric fields that are measurable at the surface of the body. These fields are used in medical diagnosis.

Essentially electricity gives rise to magnetism and magnetism gives rise to electricity. So in biological systems, bioelectricity gives rise to biomagnetism and biomagnetism gives rise to bioelectricity

Biomagnetic fields: The magnetic permeabilities of the various tissues are all about the same, approximately 1, as in a vacuum.

Bioelectric fields: The electrical resistances of the various tissues vary by a factor of about 30. Bioelectric fields generated within the body take the paths of least electrical resistance, so the patterns measured at the body surface are intricate and much more difficult to interpret.

MEASURING THE MEASURABLE HUMAN ENERGY FIELDS.

It has long been known that activities of cells and tissues generate electrical fields that can be detected on the skin surface. But the laws of physics demand that any electrical current generates a corresponding magnetic field in the surrounding space. Since these fields were too tiny to detect, biologists assumed they could have no physiological significance.

This picture began to change in 1963. Gerhard Baule and Richard McFee of the Department of Electrical Engineering, Syracuse University, Syracuse, NY detected the biomagnetic field projected from the human heart.

Subsequently, it has been discovered that all tissues and organs produce specific magnetic pulsations. Collectively these magnetics pulsations have come to be known as biomagnetic fields. The traditional electrical recordings, such as the electrocardiogram and electroencephalogram, are now being complemented by biomagnetic recordings, called magnetocardiograms and magnetoencephalograms.

For various reasons, mapping the magnetic fields in the space around the body often provides a more accurate indication of physiology and pathology than traditional electrical measurements.

PATHOLOGY ALTERS THE BIOMAGNETIC FIELD

In the 1920's and 1930's, a distinguished researcher at Yale University School of Medicine, Harold Saxon Burr, suggested that diseases could be detected in the energy field of the body before physical symptoms appear. Moreover, Burr was convinced that diseases could be prevented by altering the energy field. It also means that dramatically altering the electromagnetic field of the body can cause disease.

LATEST RESEARCH

Measurements of the biomagnetics of the heart and the brain led to a veritable explosion of research into biomagnetics. It turns out that biomagnetic fields are often more indicative of events taking place within the body than are electrical measurements at the skin surface. Every muscle produces magnetic pulses when it contracts. The larger muscles produce larger fields the smaller muscles, such as those that move and focus the eyes, produce tiny fields. This may be of interest to movement therapists, because we know that any movement from any part of the body is broadcast into the space around the body as a "precise biomagnetic signature of movement."

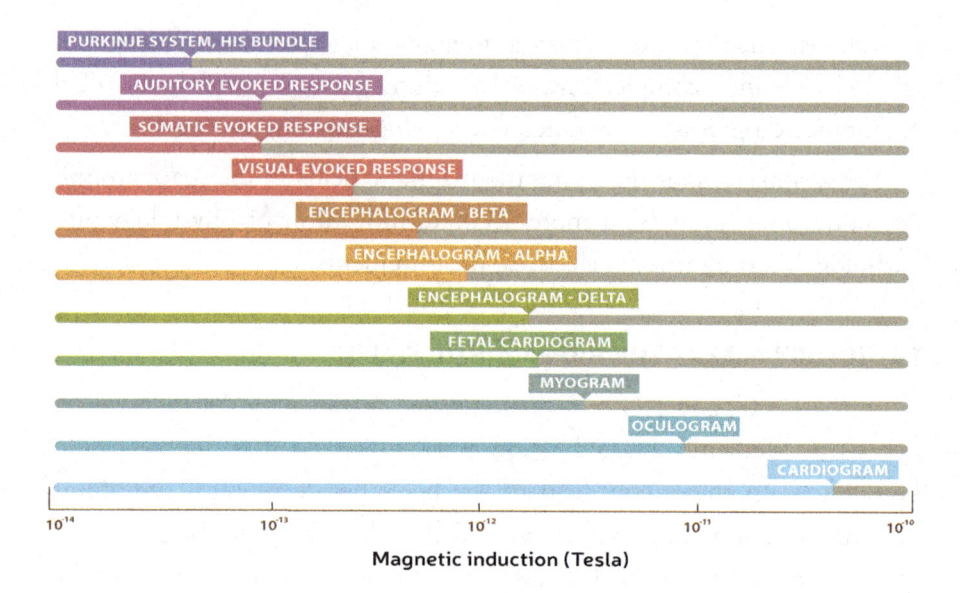

Fig. 1: Relative strengths of the various biomagnetic fields measured in the spaces around the human body (Williamson & Kaufman, 1981).

Note that all the above fields are many times smaller then EMR emitted by Regular GSM mobile phones and PDAs emit both pulsed radio waves (from the antenna) and ELF (from the battery circuits). Is it any wonder than when the magnitude of the radio waves is thousand to a million times more then the naturally occurring fields of the human body, that the body goes into shock on exposure?

HOW CHEMISTS THINK ABOUT WATER

The nature of liquid water and how the H_2O molecules within it are organized and interact are questions that have attracted the interest of chemists for many years. There is probably no liquid that has received more intensive study, and there is now a huge literature on this subject.

The following facts are well established:
- H_2O molecules attract each other through the special type of dipole-dipole interaction known as hydrogen bonding

- a hydrogen-bonded cluster in which four H_2Os are located at the corners of an imaginary tetrahedron is an especially favorable (low-potential energy) configuration, but...
- the molecules undergo rapid thermal motions on a time scale of picoseconds (10^{-12} second), so the lifetime of any specific clustered configuration will be fleetingly brief.

A variety of techniques have been used to probe the microscopic structure of water. These including infrared absorption, neutron scattering, and nuclear magnetic resonance. The information garnered from these experiments and from theoretical calculations has led to the development of around twenty "models" that attempt to explain the structure and behavior of water. More recently, computer simulations of various kinds have been employed to explore how well these models are able to predict the observed physical properties of water.

ELECTRIC FIELDS AND WATER

Water, being dipolar, can be partly aligned by an electric field and this may be easily shown by the movement of a stream of water by an electrostatic source. Very high field strengths (5×10^9 V m^{-1}) are required to reorient water in ice such that freezing is inhibited.

Even partial alignment of the water molecules with the electric field will cause pre-existing hydrogen bonding to become bent or broken. The balance between hydrogen bonding and van der Waals attractions is thus biased towards van der Waals attractions giving rise to less cyclic hydrogen bonded clustering.

MAGNETIC FIELDS AND WATER

Water is diamagnetic and may be levitated in very high magnetic fields (10 T, cf. Earth's magnetic field 30 µT)]. Lower magnetic fields (0.2 T) have been shown, in simulations, to increase the number of monomer water molecules but, rather surprisingly, they increase the tetrahedrality at the same time.

They may also assist clathrate formation. The increase in refractive index with magnetic field has been attributed to increased hydrogen bond strength. These effects are consistent with the magnetic fields weakening the van der Waals bonding between the water molecules and the water molecules being more tightly bound, due to the magnetic field reducing the thermal motion of the inherent charges by generating dampening forces.

There is a fine balance between the conflicting hydrogen bonding and non-bonded interactions in water clusters. Any such weakening of the van der Waals attraction leads to a further strengthening of the hydrogen bonding and greater cyclic hydrogen bonded clustering.

This effect of the magnetic field on the hydrogen bonding has been further supported by the rise in the melting point of H_2O (5.6 mK at 6 T) and D_2O (21.8 mK at 6 T). This indicates greater ordering (lower entropy) in the liquid water within a magnetic field. Far greater effects on contact angle and Raman bands have been shown to occur using strong magnetic fields (6 T) when the water contains dissolved oxygen (but not without the paramagnetic oxygen), indicating effects due to greater clathrate-type water formation.

<u>Impact of electromagnetic fields on water</u>

In addition to the breakage of hydrogen bonds electromagnetic fields may perturb in the gas/liquid interface and produce reactive oxygen species. Changes in hydrogen bonding may affect carbon dioxide hydration resulting in pH changes. Thus the role of dissolved gas in water chemistry is likely to be more important than commonly realized [particularly as the formation of nanobubbles containing just a few hundred or less molecules of gas], the stability of larger bubbles (~300 nm diameter) detected by light scattering and nanobubble coating of hydrophobic surfaces have all been recently described.

Reinforcement of this view comes from the effect of magnetized water on ceramic manufacture and out-gassing experiments. This effect that apparently result in the loss of magnetic and electromagnetic effects or photoluminescent effects. Gas accumulating at hydrophobic surfaces promotes the hydrophobic effect and low-density water formation.

The accumulated gas molecules at such hydrophobic surfaces becomes supersaturating when electromagnetic effects disrupt this surface low-density water. An interesting (and possibly related) 'memory of water' phenomena is the effect of water, previously exposed to weak electromagnetic signals, on the distinctive patterns and handedness of colonies of certain bacteria.

ELECTRODYNAMIC PROPERTIES OF WATER

There is scientific proof that extremely low frequency electromagnetic field can dramatically affect the dielectric permittivity and electrical conductivity of water and water based solutions. The scientists of Novocontrol Technologies GmbH & Co. KG (http://www.novocontrol.com) provide the following results for measurements of electrodynamic characteristics of water when the body of water is exposed to the wide range of electromagnetic oscillations

In the range of low frequencies of 0.1 – 1000 Hz the relative dielectric permittivity of water increases from its regular value of 80 up to 108, and electrical conductivity decreases up to 10 times. These facts confirm that water as a subject to applied EMF of extremely low frequency range undergoes molecular structural modifications. It is reasonable also to admit that these structural changes can affect the electrodynamic characteristics of water in the range of RF frequencies as well. The relative dielectric permittivity of water significantly increases from 80 up to 108 and electrical conductivity of water samples decreases up to 10 times in the frequency range of 0.1 .– 1000 Hz. Measurements were conducted on the samples of deionized and tap water in measurement units with different size (length and diameter) at 20°C.

Under the influence of applied EMF polar molecules tend to align themselves with the field. Although water has polar molecules, its hydrogen bonding network tends to oppose this alignment. The level to which a substance does this is called its dielectric permittivity.

Dependent on the frequency of applied EMF the dipole may move in time to the field, lag behind it or remain apparently unaffected. The ease of the movement depends on the viscosity and the mobility of the electron clouds. In the wide range of EMF frequencies lower than GHz frequency level (corresponding to microwave thermal effect) the water dipoles move in time to the field.

ELECTRODYNAMIC CHARACTERISTICS OF WATER

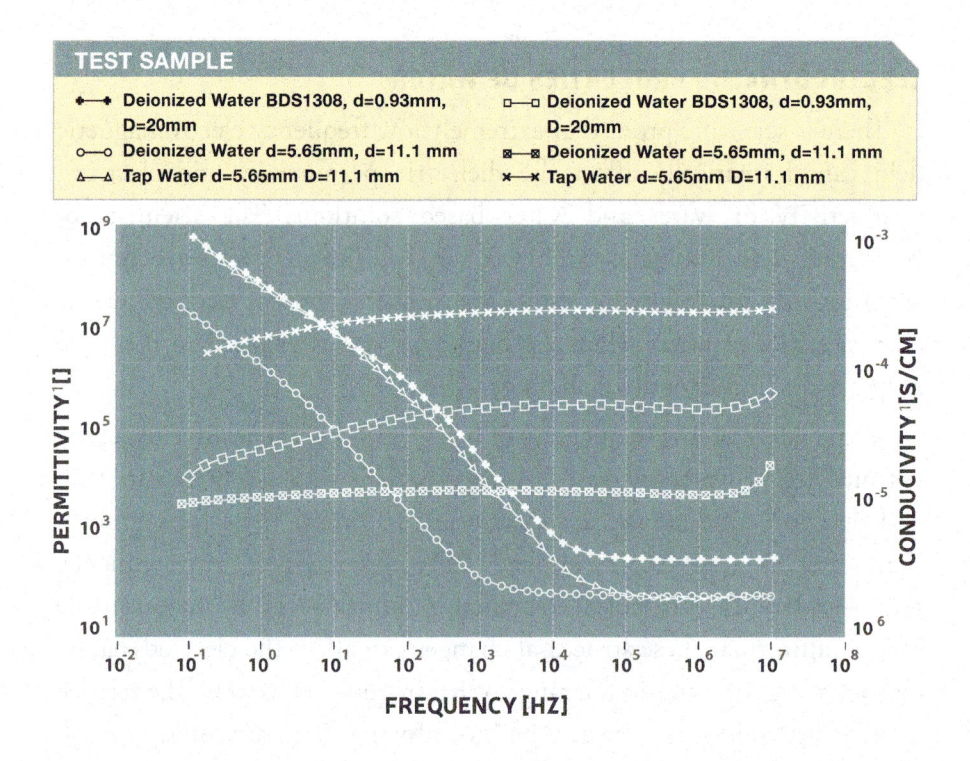

Fig. 2: Electrodynamic Characteristics of Water

In the range of extremely low frequency of 0.1 – 1000 Hz (corresponding to the extremely low velocity of movement) the dynamic viscosity of water and the resistance of water dipoles to the alignment (dielectric permittivity) are extremely high (up to 108 at 0.1 Hz) due to hydrogen bonding between molecules (molecular coupling).

In this extremely low range of frequency of applied EMF the water dipoles are able to move in time with low frequency electromagnetic field. As a result they can form multilayer molecular formations which oscillate in accordance with applied low frequency EMF.

In the higher frequency range of kHz to GHz generated by RF phones the reorientation process may be modeled using a "wait and switch" process where the water molecule has to wait until favorable orientation of neighboring molecules occurs and then the hydrogen bonds switch to new molecules. This range of frequencies of RF phones is related with the ease of the movement of water dipoles resulting in chaotic Brownian movement of water molecules.

In the process of Brownian movement water molecules located in close proximity to each other develop the "friction effect". That results in an increase of the level of absorption of EMF energy emitted by RF phone and in the generation of heat (called "dielectric loss."). This process creates free radicals in cellular tissue leading to radiation damage.

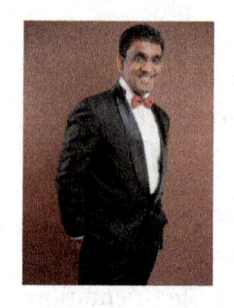

DR SUNDARDAS D. ANNAMALAY

Naturopathic Physician, Acupuncturist, Homoeopath, Clinical Nutritionist, NLP Trainer, Clinical Hypnotherapist

Professor, YINS Colleges Worldwide, Researcher

International Speaker, Coach, Consultant

Asia's leading Naturopathic and Wellness Medicine Expert

As Seen on:

Dr Sundardas is the leading Naturopathic Physician practicing for the last 25 years in Singapore. His clinical interests include children's learning disabilities (ADD/ADHD, Autism, Infections), Allergies, Women's Health Concerns, Musculoskeletal Pain and Healthy Aging. He is currently Professor of Naturopathic Medicine to the Youngson Institute of Natural Science (Australia) and runs a busy practice in Singapore.

He has been a visiting professor to the Open International University for Complementary Medicine (Sri Lanka). He is CEO of Natural Therapies Research Centre Pte Ltd and Sundardas Naturopathic Clinic, Asia's first ISO registered Complementary Medicine Clinic.

Dr Sundardas is a Diplomate certified by the American Board of Anti-Aging Health Professionals. He is a member of the American Academy of Anti-Aging Medicine. He is also a registered Naturopathic Physician with the Naturopathic Practitioners' Association (Australia). Dr Sundardas is a Fellow of the Faculty, University of Natural Medicine (Nevada).

His client list includes some of the leading Asian actresses and entertainers, diplomats ambassadors, political figures from around the region, some of the richest families in Asia, the Middle East, India and members of the royalty from Malaysia. It also includes executives from organizations like Goldman and Sachs, IBM, SIMEX ,FOREX and other companies like Batey Ads, Business Trends, NTUC, as well as local government bodies like Ministry of Defence and Ministry of Education.

He is a speaker, trainer and consultant to several health related companies. He has also won awards both local and foreign as well as for his contributions ranging from "Associate of the Teachers' Network" awarded by the Ministry of Education (Singapore) to Dr Yudvir Singh Memorial Award (India) to being listed in the "Who's Who of Intellectuals").

Dr Sundardas is a well respected media personality and has been interviewed by TCS, CNBC Asia and BBC World. He has also been interviewed many times on radio and ran a regular talk-show on Complementary Medicine in Singapore. He has seen more than 15, 000 individuals both in his clinic and his seminars. His print, radio and television appearances over the last 25 years in Singapore, Malaysia, Thailand, Hongkong and India would have influenced a few hundred thousand more.

He is author of the following books:
- Asian Woman's Guide to Health, Beauty and Vitality (EPB)
- Awakening the genius in your child (EPB)
- Maximise your child's emotional intelligence (EPB)
- The Science of Healing Water (Times)
- Everywomen's Guide to Vitality and Anti-Aging
- Water –The Untold Story
- "Pindlahr Techniques" by Dr William Gordon Youngson and Dr Sundardas D Annamalay (Youngson (WG) and Associates)
- Sex Education for Autism (editor)
- Recipe for Perfect Health
- The Art and Science of Vibrational Medicine
- "Out Front" co-author , Amazon Best Seller 2012

E-Books

- Allergy Free with Naturopathy
- Pain Free with Naturopathy
- The Way of the Gentle Warrior + workbook
- The Science behind the secret + workbook
- Tools for manifestation + workbook
- Tapping your fears, phobias and allergies away - Introduction to Fractal Field Therapy + workbook
- Turning on your Success Blueprint – Success Permission + workbook

www.AgainstRadiation.com

CPSIA information can be obtained
at www.ICGtesting.com
Printed in the USA
FSOW03n1316261216
28856FS